Gottfried Bammes • The Artist's Guide to Human Anatomy

Gottfried Bammes

The Artist's Guide to Human Anatomy

An illustrated reference to drawing humans including work by amateur artists, art teachers and students

CHARTWELL BOOKS. INC.

Translation: Judith Hayward in association with First Edition
Translations Ltd, Cambridge

This edition published in the USA in 1994 by Chartwell Books
Inc, a Division of Book Sales Inc, 114 Northfield Avenue, Raritan
Center, Edison, N.J. 08818

Printed in 1994 in Spain

ISBN 0-7858-0054-9

List of contents

Introduction

This book presents the reader with almost 200 study drawings which demonstrate that a confident working understanding of anatomy can be an invaluable tool for the artist. The drawings were produced over a period of four semesters of weekly three-hour lessons. The standard achieved after one year of study was tested by an intermediate three-part exam lasting four hours partly based on visual imagination, and the second year was concluded by a drawing project, again in three parts and again partly based on visualization.

For the tests at the end of the second semester (first year), students were allowed a short time with a model to study his or her pose and analyze the constructional skeletal forms, the groups of functional muscles in action and the living form of the leg (see fig. 63). The final exam at the end of the fourth semester (second year) followed the same procedure – extended to the whole figure – with a special bias (see fig. 162) relevant to the student's area of study. This book also contains examples of work produced by students attending Bammes intensive courses at Zurich and Salzburg.

What training and educational objectives, beyond anatomical knowledge, do the drawings reproduced in this book demonstrate?

When faced with a nude model, beginners – as well as feeling embarrassed – are usually confused about where to begin. The information about proportion provides important basic guidance suggesting that the figure be approached as an articulated entity in which the parts relate to the whole (figs. 1–13). The forms of the human body have varying characteristics and interactions. Values relating to impression and degrees of visual weight are discovered, to do with the main forms and the subsidiary, intermediate and transitional forms that accompany them. We work out a hierarchy of forms – not only in the context of studies of proportion – and the connections that exist within it.

Anatomy for artists aims to highlight what unites rather than what separates, and it is only to work toward this end that we undertake studies of individual parts.

Another priority for becoming fluent in figure drawing is familiarity with the way the figure at rest and in motion invariably conforms to certain rules (figs. 17–41). The rules learned when studying proportion can easily be extended and applied to the body. It must be realized that a standing or sitting position (figs. 13–16 and 28–34) is not just a question of mechanics – there is always a psychological aspect as well. The person drawing must always empathize with the model's gestures, sharing in the experience and endurance of tension and its release. This book provides introductory examples and methods of procedure that are practicable even for beginners.

Of course these first life drawings and what they can achieve in terms of a conceptual complexity have their limits. In order to understand how the human body functions, students must also familiarize themselves with the structure of its individual parts. Only close studies supported by analytical methods enable students to convey form and function as they interact. If, for example, you want to understand the essence of what happens to the form of the knee when it bends (figs 57, 59–65), you must have a thorough knowledge of the construction of the skeleton of the knee as regards form, components, structure and mechanics. The same is true of the torso (figs 83–95), pelvis (figs 58–60), rib cage and the mechanics of the spinal column. The behavior of the soft, fleshy forms can then be inferred. Thus, we have to construct the figure from depth if we want more than a clever but facile result. That is why the forms of the human framework must be simplified as far as possible, and why I always put greater emphasis on the abstractive skeleton forms than on the musculature (figs 79, 80, 82, 66, 102–109).

Of course, once individual parts of the body are known to the student (and can be drawn from a mental image of them), they have to be fitted back into the whole. Consequently the sequence of drawings contained in this book demonstrates that all graphic studies of parts of the body or the body as a whole must first be supported by the ability to draw the constructional forms. They are a distillation of form, conveying information regarding what is or is not possible in functional and plastic terms. Conveying how the organism responds to the demands made upon it to perform also becomes the purpose of the drawing.

If you consider depicting forms of the body to be a kind of structural drawing, then it is even more important to have an architectural understanding of the live appearance of the body in order to portray it (figs 152, 87, 153–159). It is a question of working out the structural interplay between supporting and supported forms, between the relatively constant framework forms that provide stability and the changeable, pendulous forms of the soft parts of the body. This applies not only to the structural processes of the figure as a whole, but also to the functional ones. In other words the body conceived as a building must evoke in the person drawing it the main elements of a building, the solid core shapes and the container shapes, and at the same time it must be possible to see through these so that each individual item is assigned to its proper place. Constructional form determines appearance and is a component of essential form.

Drawing form constructionally may be a valuable training exercise but there are other prospects and purposes implicit in it. The behavior of the soft, fleshy forms is a consequence of the behavior of the framework. In architectonic form – which Adolf Hildebrand saw as a form of artistic nature study – this consequence does not produce showy muscular figures, nor the limp forms of dead matter which always crop up in artists' anatomy even today. Anatomy for artists conceived in this way has promoted an illustrative outer appearance of form, not essential form. Perceiving form architecturally may be seen as a special way of understanding nature, as a statement about nature. The spirit behind the concept of something constructed also favors this approach (figs 182–184, 78).

So, once more, effort is needed if the artist is to be free of total dependency on a model. When I insist that students must be able to construct a figure purely from their imagination without the presence of an actual model, they have to be such masters of their craft that they can draw the figure in its natural, human form (fig. 194). But this does not mean that from then on they can dispense with a model. Only once the fundamental facts have been fully absorbed can the unique form of each and every model – the head, arm, hand or complete figure – be fully experienced and elaborated.

The capacity for personal and subjective interpretation can come into play only once students have a command of the complete repertory of rich practical knowledge and the ability to visualize – combined with the development of the inner eye. At that point the wide field of artistic creativity opens up before them. Of course no amount of teaching can succeed in transforming the 'mortal coil' into a work of art. Everyone must embark on that journey alone, but we can take the student near to the cross-over point (figs 171, 174, 182–185, 194). The naked human body as an 'object' of life study is as great a mystery as art itself.

I see drawing human forms primarily as an organizational task. Understood in this way it can capture an important piece of vividness from our world which is becoming ever less vivid. Creating order through drawing means wresting a small bit of our world away from chaos and making it accessible to scrutiny and certainty.

Seeing with understanding and drawing with understanding may at present be regarded with suspicion as having a science-based, therefore indiscreet, relationship with the body. The veil covering a great mystery is lifted ever so slightly.

1.

Studies of proportion

The approach we adopt rules out any rigid theory of proportion with set basic measurements (modules) and an established esthetic canon. Instead we use a method that starts with the individual proportions of each quite different model. The total height of the model is structured by establishing measuring points and checking the sections marked off by them for coincidences or similarities. In practical terms this means that we first establish the overall height of the figure from the sole of the foot to the crown of the head, draw the middle axis of the body as a connecting line and mark off on it as horizontal axes the lengths measured from the model that are alike or similar (height proportioning). The geometric center of a woman's body is certain to coincide with the position of the pubic bone, i.e. the length of the upper part of the body corresponds with that of the lower part.

After the first basic proportioning, other reference points in the upper and lower parts of the body are investigated. For example we find the position of the nipples (figs 1–6) a little below the top quarter of the body, while the lower edge of the kneecap (coinciding with the interarticular space) is in the bottom quarter, which means that the length of the upper leg is equal to that of the lower leg plus the foot. Height proportioning is further refined as indicated. The height of the head (chin to the top of the skull), the length of the neck, the position of the navel and the waist, the length of the knee as an intermediary form between the upper and lower leg and the height of the inner ankle are also noted.

Once it has been established, the head height (H) is an appropriate measuring unit or module for determining other measurements and assessing how far an individual model conforms to the canons of proportion. Figs 1–6 show that the models used conformed to the 8-H canon in proportion.

Measurements that are less than one head height can be expressed as fractions of it: neck-length c. $\frac{1}{2} - \frac{1}{3}$H, height of ankle $\frac{1}{3}$H, and so on. As well as being referred to the H module all measurements should be compared with one another.

Once the height proportions have been established, a similar process is carried out to ascertain the widths of the horizontal axes of the body. Typically the widest measurement in a female body is across the broadest part of the hips, often exactly 2H, while the narrower measurement between the edges of the shoulders is often approximately $1\frac{1}{2}$H.

Since the widest body measurement on a woman (across the hips) does not exceed 2H, a rectangle 2H wide by 8H high can be blocked in as a check on the figure; subsequent freer proportion studies can then be sketched in to provide a broad overall view. If the widest points of the horizontal measurements are joined, a simple proportioned figure consisting of geometrized forms emerges. This simple guide to proportion is not a rigid dogma, and can be applied to every model.

It is extremely important to work out the rough, broad forms of the sections of the body (e.g. the rectangle of the torso, the trapezoid of the hips, the intermediate shape of the stomach, the ovoid shape of the head, etc):

- Simple forms are easy to remember and eye-catching.
- The fact that simple forms are easy to visualize helps you to understand functional processes (figs. 18–22). Only once they have been understood will you be able to convey the behavior of the forms of the soft, fleshy parts of the body clearly and convincingly, e.g. compression and stretching.

After drawing up the geometrized proportioned figure, two further stages are necessary:

- Making sense of the contours and formal accents by drawing in extremely simplified skeleton forms (fig 4, blue lines). These represent the basis for the formal structure of each section of the body and for understanding organic and functional processes (figs 17–24).
- Filling in characteristic subsidiary forms (e.g. breasts, layers of fat on the hips, flanks and knees, figs 3, 5) on the outline figure.

It is a good idea to master all these stages using broad lines to avoid creating indeterminate images and to achieve precision of form.

Only now is it appropriate to undertake three-dimensional studies of the body in both front and back views (figs 6, 8):

- Study the spatial gradations and draw in the forms of the body nearest to the observer and those around them using denser or lighter shading.

Drawings that have an almost pictorial quality (figs 9, 16) can be achieved by producing a composite whole figure (these are not silhouette cut-outs!).

New problems relating to proportion arise when we come to the profile. As well as the procedures that have already been covered we have to consider the rhythmic pattern of the body (figs 10–15):

- The figure is constructed around the structural line that runs through the opening of the ear and the joints of the shoulder, elbow, hip, knee and ankle like a plumb-line.
- The slanting axes formed on the side of the stomach by the volumes represented by the rib cage and pelvis create an obtuse angle (so there must be no vertical stratification!).
- The buttocks, front of the thighs and the calves along with the upper body form shapes that project alternately to back and front.

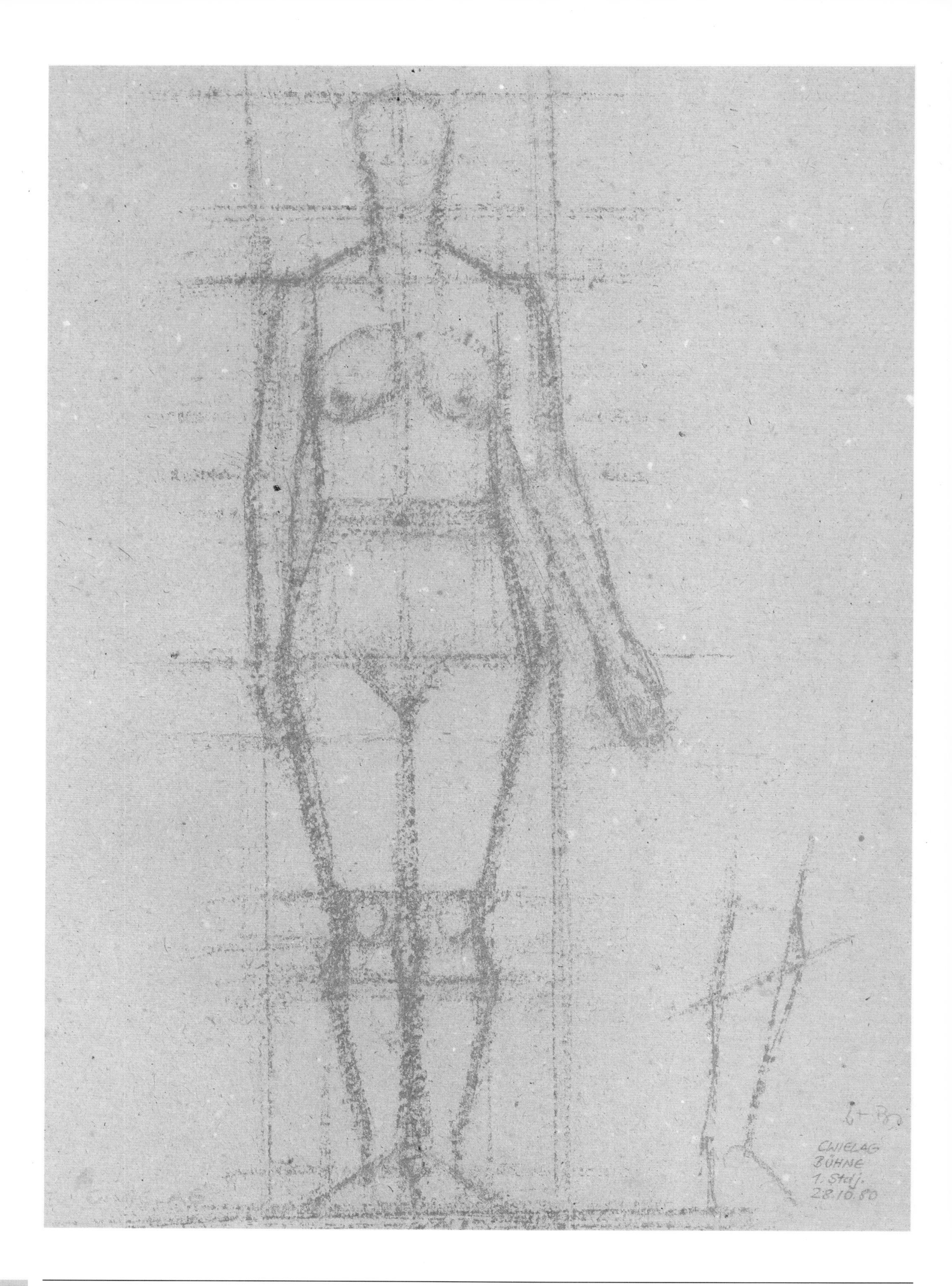
CWIELAG
BÜHNE
1. Stdj.
28.10.80

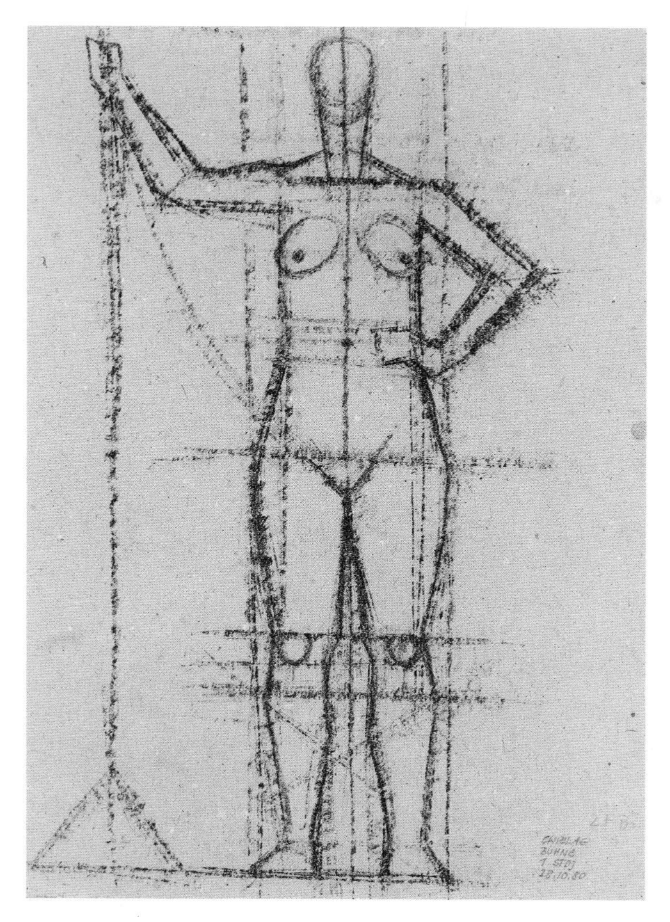

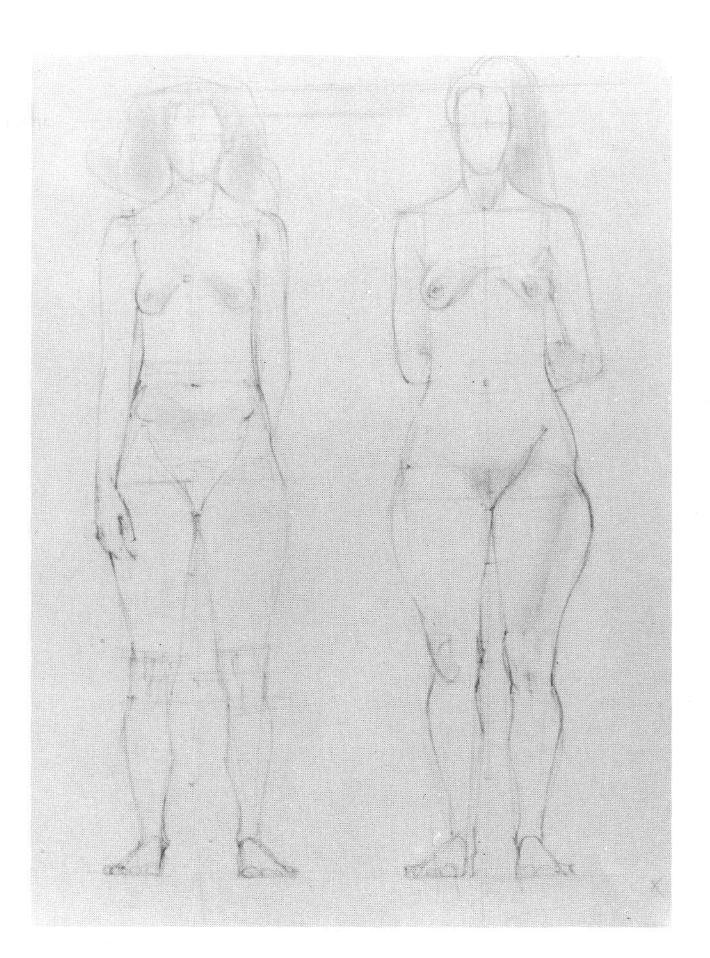

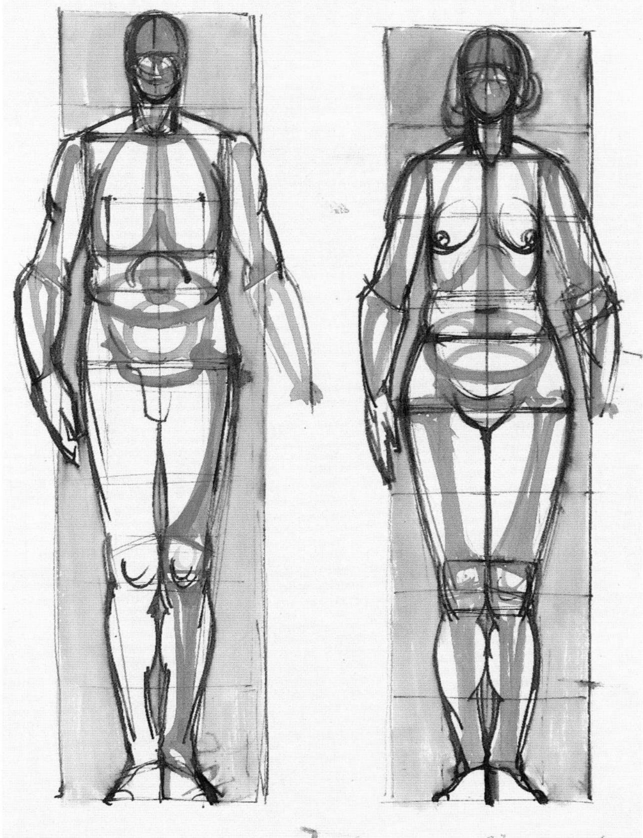

1 GEOMETRICALLY SIMPLIFIED PROPORTIONED FIGURE
This study is executed in red chalk crayon using long unbroken lines so as not to get lost in anatomical details. The procedure for marking off height proportioning can be recognized from the lightly drawn horizontal axes.
Student of stage design, first semester

2 GEOMETRIZED PROPORTIONED FIGURE, IN A VARIATION OF THE STANDING POSITION
The new directions of the limbs and the shapes of the intermediate spaces enclosed by axes produce useful check-points in the drawing of the figure.
Student of stage design, first semester

3 DRAWING IN SUBSIDIARY, INTERMEDIATE AND TRANSITIONAL FORMS ON THE PRELIMINARY PROPORTIONED FIGURE
This phase uses what has already been learnt to introduce a stronger suggestion of individualized live appearance.
From the Salzburg intensive course, 1988

4 MAKING SENSE OF THE OUTLINE OF THE FIGURE BY INSERTING SIMPLIFIED SKELETON SHAPES
The geometrized shapes should not be adopted by the student as schematic forms without any further thought. The 'corners', or accents, of the outline shape are given a primary anatomical and organic justification. Student of painting/graphic art, first semester

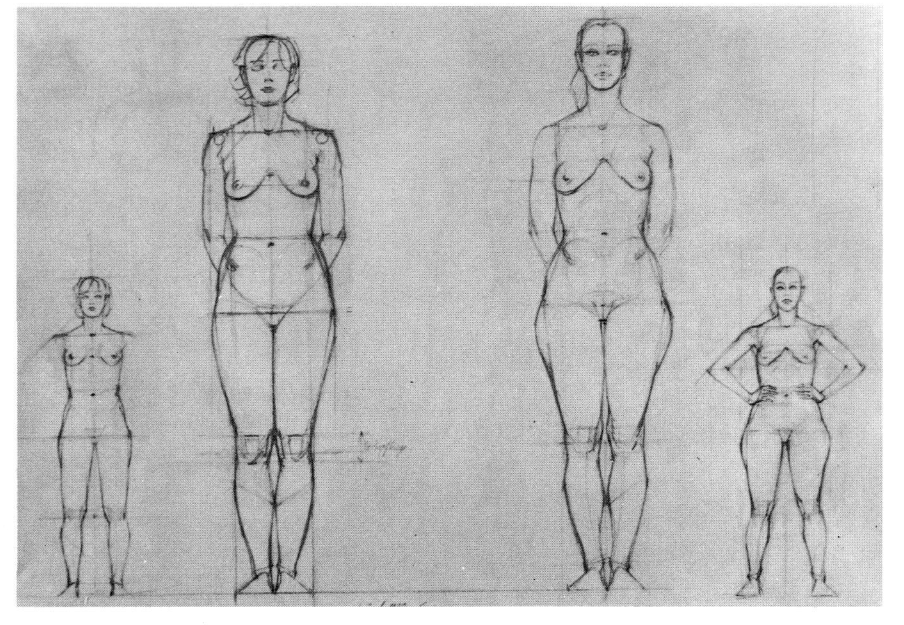

5 COMPLETED STUDIES OF THE PHYSICAL APPEARANCE OF TWO DIFFERENT MODELS

It is clear from the figures that the 'flexible' method of recording proportions has nothing to do with stereotyped 'proportion templates'. Straight and rounded contour lines join up.
From the Salzburg intensive course, 1988

6 EXTENDING THE STUDY OF PROPORTION BY EXAMINING THE BODY FROM A THREE-DIMENSIONAL VIEWPOINT

The parts of the body that protrude farthest are indicated using different weights of shading. The proportional construction of the body is now primarily based on estimation.
From the Salzburg intensive course, 1988

7 STUDY OF ALTERED PROPORTIONAL RELATIONSHIPS IN BACK VIEW

The horizontal axis at the middle of the body is still the basic guideline. The horizontal crease made by the buttocks is below that level. The line of the shoulder bones runs down from the tip of the shoulder.
From the Salzburg intensive course, 1988

8 STUDY OF PROPORTION WITH MORE DIFFERENTIATION

It is not only external forms that are differentiated; internal shapes are also perceived three-dimensionally in very subtle gradations relating to spatial rather than anatomical factors. Light, open areas indicate concavities.
From the Salzburg intensive course, 1988

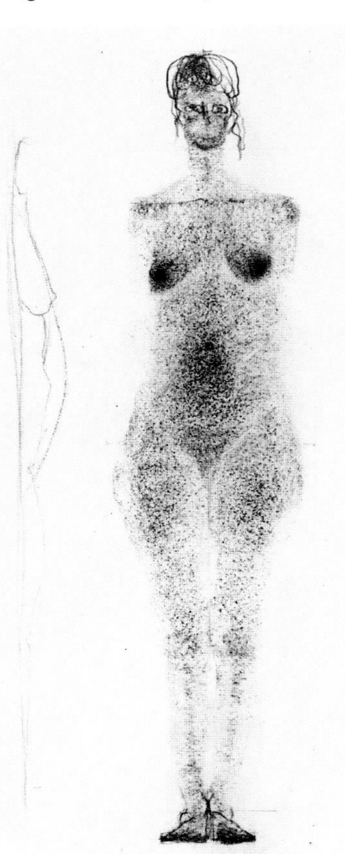

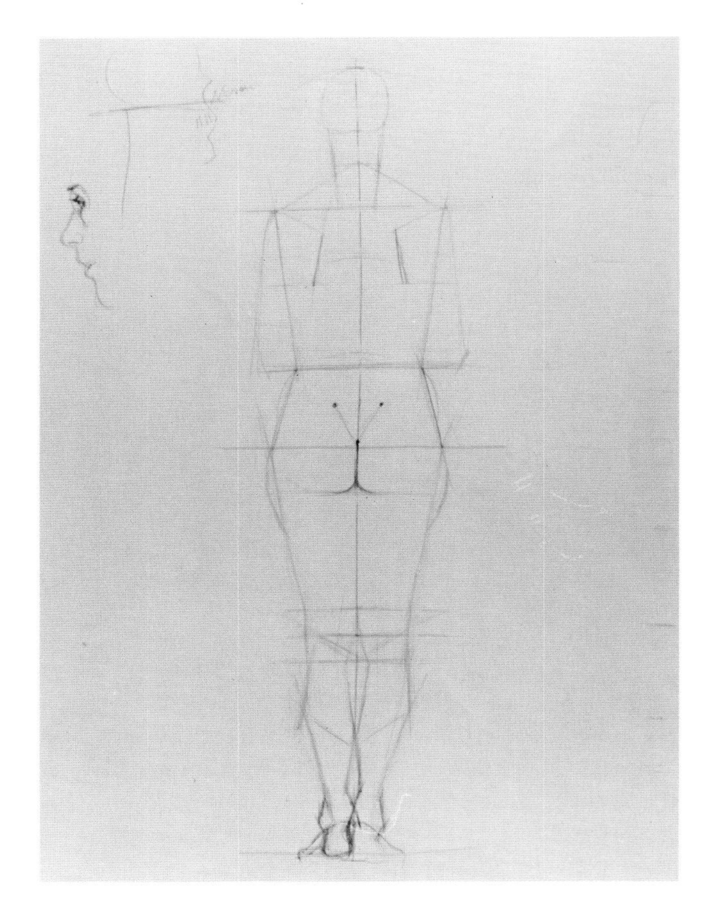

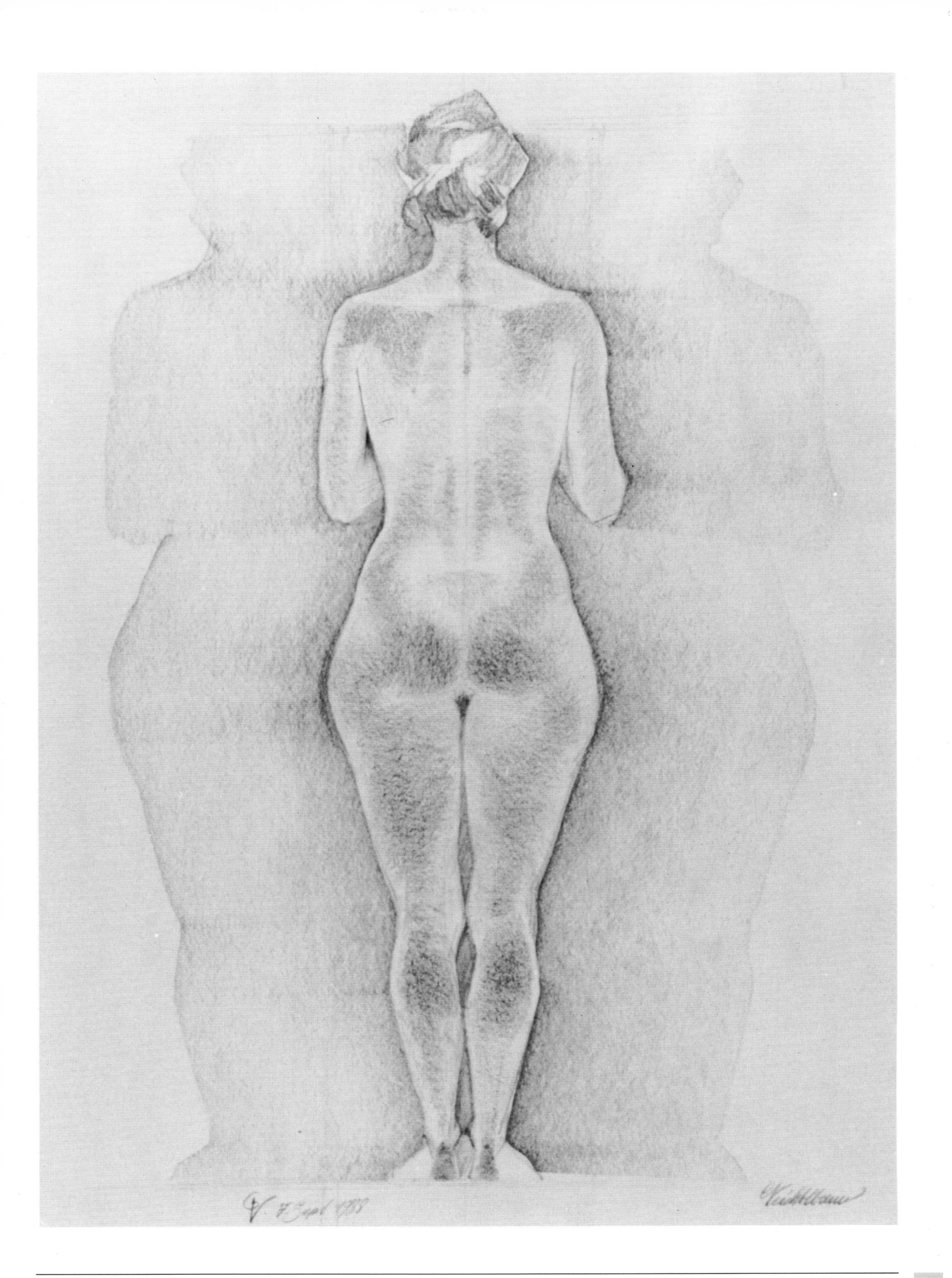

9 PROPORTIONALITY LINKED WITH SURROUNDING SPACE

The two figures sketched in this study of proportion are placed in spatial relation to one another (the intervening space) and the surrounding space (the floor surface and the partition). The kinetic hatching is no longer tied to the linear silhouette.

From a Bammes course at the Schule für Gestaltung, Zurich

10 RHYTHMIC PATTERN OF THE BODY IN PROFILE

Awareness of proportional factors is linked with the balance of the masses lying one under another and projecting both forward and backward. Red chalk applied lengthwise gives the direction of the masses as they first strike the viewer, not relying on mechanical hatching.
From the Salzburg intensive course, 1988

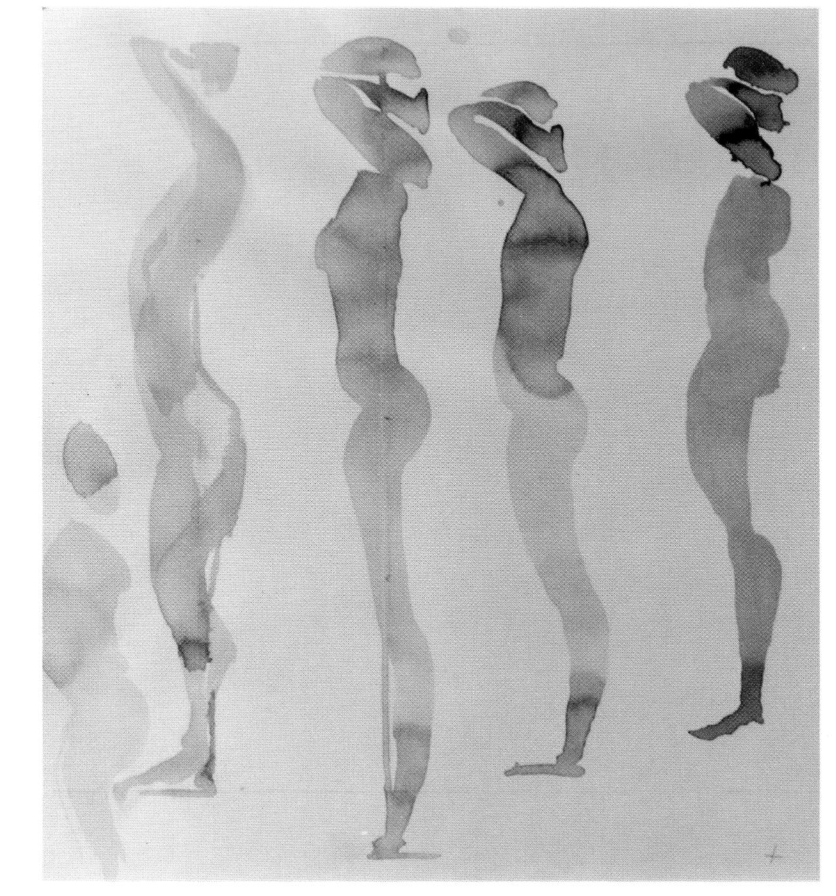

11 RHYTHMIC PATTERN OF THE BODY AS A COHERENT FLOW

To turn the sequence of forms into something really fluid, use a well-primed watercolor brush and trace the contours of the body, varying the pressure used. The figure materializes in a matter of seconds – the brush is not lifted and there has been no previous calculation of proportions.
From the Salzburg intensive course, 1988

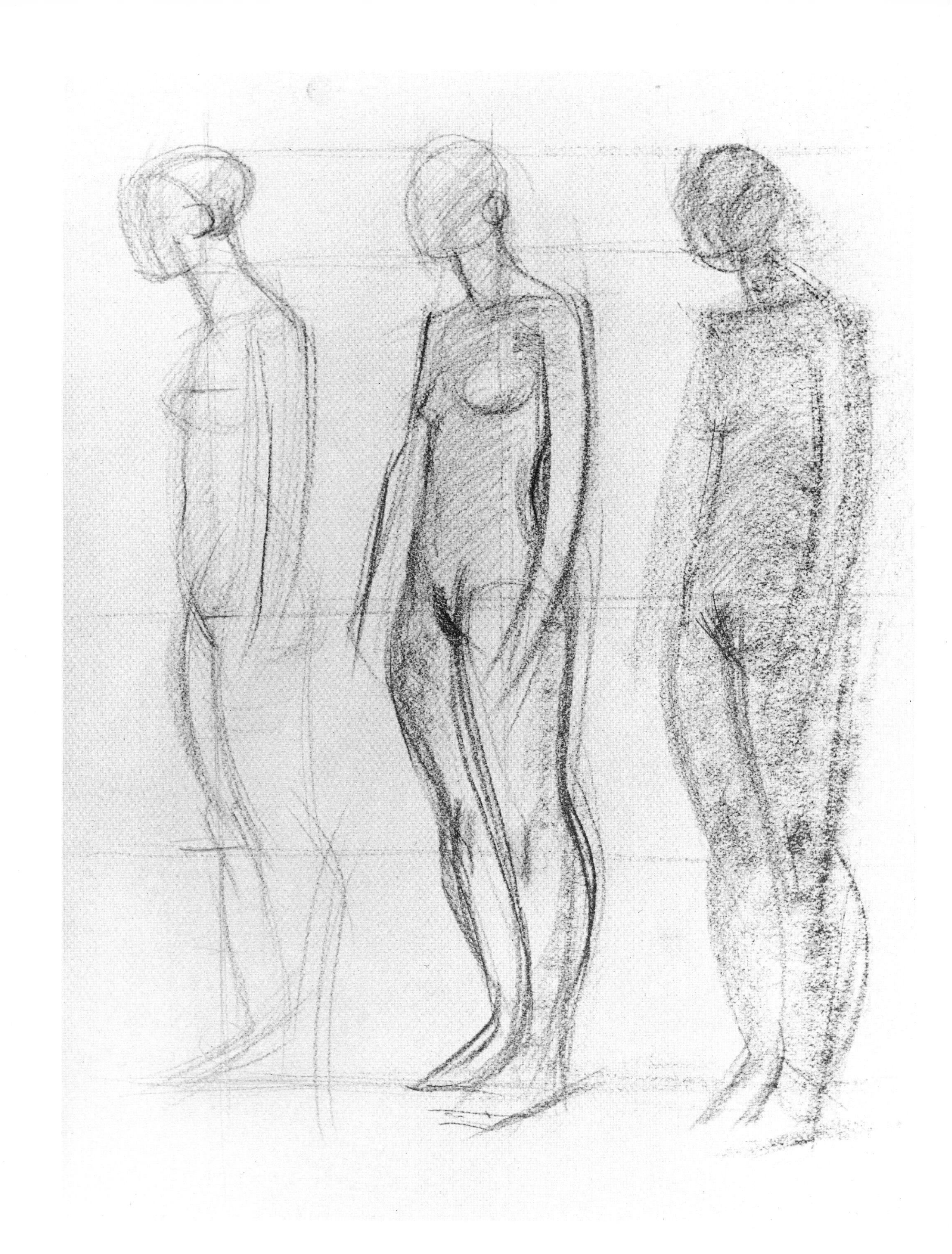

12 RHYTHM OF THE FIGURE IN THREE-QUARTER VIEW

As well as emphasizing the curves, it is necessary here to pay attention to spatial considerations such as the falling axis of the shoulder-line and the rising axis of the surface on which the model is standing. Rough proportioning guidelines can still be seen.
From a Bammes course at the Schule für Gestaltung, Zurich

13 INTEGRATING SPATIAL FACTORS INTO THE RHYTHMIC PATTERN OF THE FIGURE

A bristle brush half-filled with opaque paint is a good medium for producing a profile that can be executed quickly, and by using stronger or more muted shading you can convey a sense of depth of perspective.
From the Salzburg intensive course, 1988

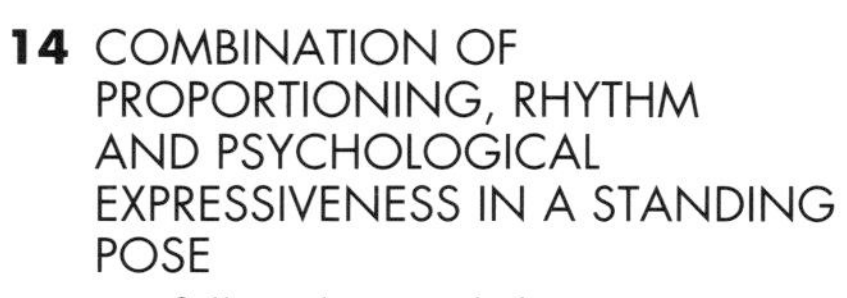

14 COMBINATION OF PROPORTIONING, RHYTHM AND PSYCHOLOGICAL EXPRESSIVENESS IN A STANDING POSE

You can follow the rapid change in a standing pose in profile by using an adaptable medium, in this case powdered sepia applied with the finger. The loose forms can be stabilized with a few contour lines. The psychological state of a model is evident even when he or she is working, and this should not be overlooked.
From the Salzburg intensive course, 1988

15 RHYTHMIC PATTERN OF THE BODY CONVEYED BY LINE ONLY

Some artists are able to give rhythmic definition to the human form using linear flow only. This study of two nudes shows the requisite mastery of outer and inner closeness, of links and breaks between the bodies.

From a Bammes course at the Schule für Gestaltung, Zurich

16 PROPORTIONAL PLANNING AND PSYCHOLOGICAL EXPRESSIVENESS IN A FRONTAL STANDING POSE

A free execution with a bristle brush supported by the broad division of the body into quarters concentrates on the gestures of the sure, self-confident pose. In this case the figure is not defined by line, but by a combination of tonal effects and open spaces.

From a Bammes course at the Schule für Gestaltung, Zurich

2.

Studies of standing and seated poses in repose

Implicit in what has been said of proportion and the accompanying studies are structural problems relating to the body's equilibrium which are particularly relevant when the model's weight is supported on one leg.

If you make cut-outs of a proportioned figure using different-colored overlaying pieces of paper (figs 17, 18), this also gives you negative outlines of the figure. You can then follow through the steps involved in structural and functional changes to the body by cutting up the positive silhouette figures into the basic geometric shapes you have already worked out; you can then simulate the processes of center of gravity shift and its implications for form:

- Over a negative silhouette of a figure standing on both feet lay the trapezoid of the hips of the positive figure so that the center of gravity which is located in the middle of the pelvis is directly above the sole of the foot of the standing leg.
- All further consequences, such as the lowering of the pelvis on the unsupported side, the slanting position of the standing leg, the compensating position of the trailing leg, the way the upper body leans over toward the side of the standing leg, etc follow from this.

This exercise makes it clear how supporting and supported forces, accents (the way the hip juts out on the side of the standing leg), compression and stretching (gaps round the intermediate form of the stomach) originate. Breaking the figure down into movable, adjustable individual cut-out parts in this way is preferable to the fashionable but ill-devised schematic methods proposed in some studios.

If you cut your proportioned figures out of foil and allow for overlapping pivotal points you can even develop an articulated mobile figure which you can use to test structural and dynamic attitudes in movement (figs 19–21). Another similar approach involves printing; make geometric shapes of parts of the body (in the form of potato cuts, linocuts or woodcuts) and use them to make prints following through what happens in different standing positions (fig. 20). Printing with blocks of movable proportioned parts of the body encourages you to identify with the expression of the movement, especially if you are one of those people who cannot immediately apply what you have learnt about proportion to different standing poses.

Closely related to the two elementary intermediate stages of cut-out exercises and printing is working with chalk: this is applied in broad lines to convey a rapid sequence of changes of pose by the model (fig. 21):

- Take a piece of chalk which is the right length to give the rectangle of the upper body in a single stroke when used on its side.
- By twisting the chalk and applying it with varying pressure you can create the forms assumed by the body in different poses: compressions, tensions, relaxations, directions and precise formal accents.

The point of this exercise would be lost if you just produced a preliminary drawing and then filled it in with mechanical hatching. Though contrapposto studies tend to be decried today as academic, a great deal can still be learnt from them in terms of understanding the interdependency of proportional principles and the functional working of the body (fig. 22). It is crucial to construct the figure according to structural rules, i.e. to convey the relationships between the position of the center of gravity and the weight-carrying surface (the sole of the foot).

The procedure is as follows:

- Decide how to break the figure down.
- Mark the weight-carrying sole and the center of gravity which is directly above it.
- Establish the direction of the standing leg, the curve described by the central axis of the body and the directions of the cross axes intersecting it at a variety of angles.

Once this structural framework has been established, incorporate the geometrized sections of the body into it before proceeding with any further differentiation of form. Rather than outlining the external forms, make the relationship between the core parts quite clear. It is important throughout these exercises that movement and changing form should emerge from our purely functional studies with a live and coherent flow, be imbued with the expression of the interdependency of the forces at work and carry conviction of the model's ability to stand and not suggest instability or toppling over!

There are many possible ways of achieving this. It may be useful – using the structural framework – to work up a figure using chalk on its side. Starting from one center (say the trapezoid of the hips) you can gradually expand upward and outward until you reach the contours (do not make a preparatory outline drawing; keep the lines broad and flat). You can use a graphite stick on its side to put in the curving main masses and directions, stabilizing the form by sketching in a very few lines (fig. 24).

A most appealing method is to use a paintbrush: the smooth, quick, shorthand-like abbreviations can grasp what is most essential (fig. 23). The brush must immediately work with volumes applied with varying weight, and not be used as a pencil to provide a preparatory drawing.

Try always to experiment with expressive exaggeration to capture the essence, while working in pen and ink or combined techniques (figs 25–27) too. When drawing you must constantly be on the look-out for changes, and this applies equally to seated poses (figs 28–34). The model should help students grasp the expressiveness of a pose by rapid changes of position which will force them to observe very quickly and prevent them from getting bogged down in anatomical detail.

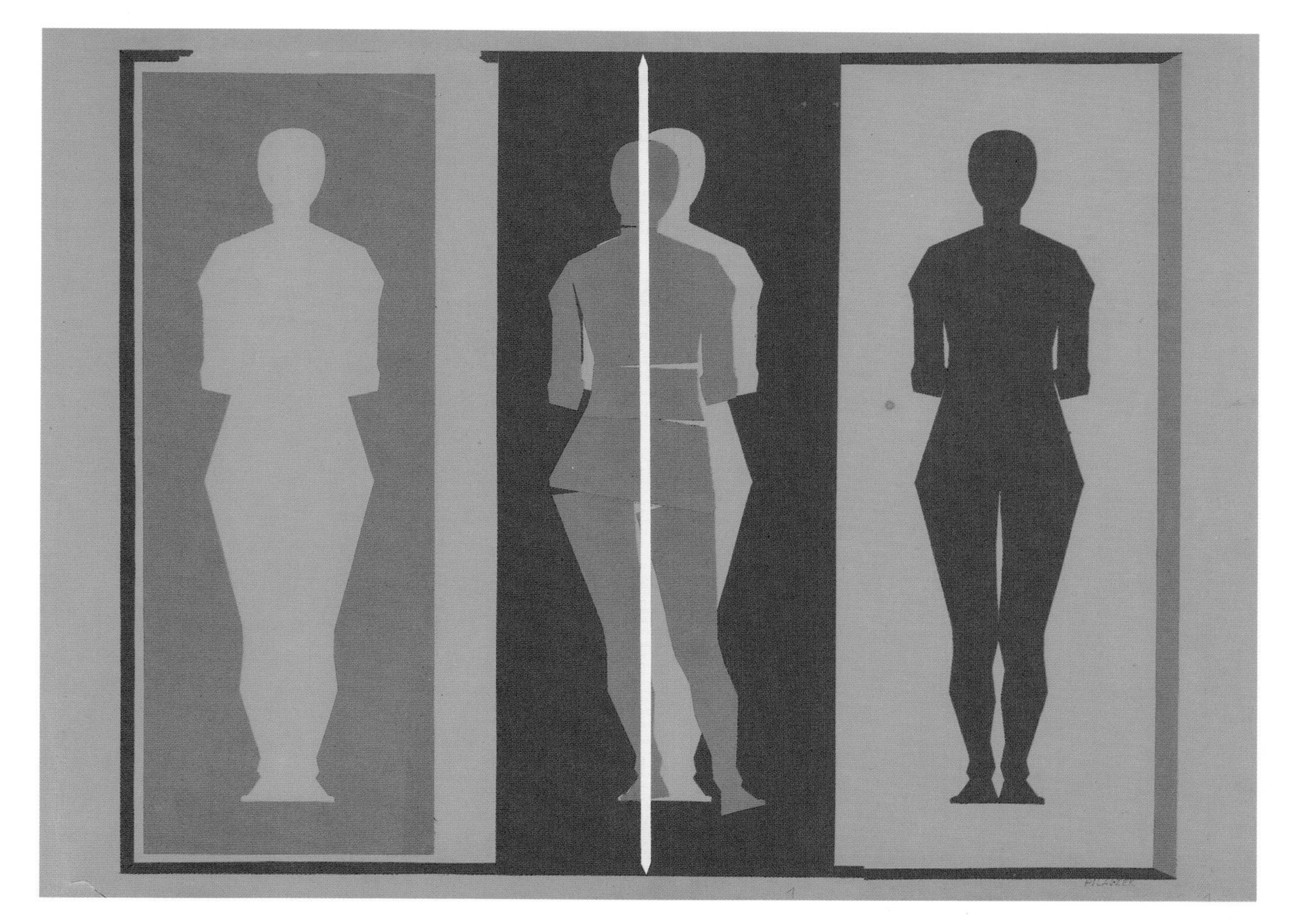

17 CONTRAPPOSTO POSE BUILT UP FROM PROPORTIONED CUT-OUTS

A lot can be learnt from understanding the many regular changes of form that occur through observing the mutual relationship of the center of gravity, the gravity line (in white) and the sole of the supporting foot. The disposition of forms is based on preserving the balance when the weight is supported on one side by the sole of the supporting foot.
Student of set painting, first semester

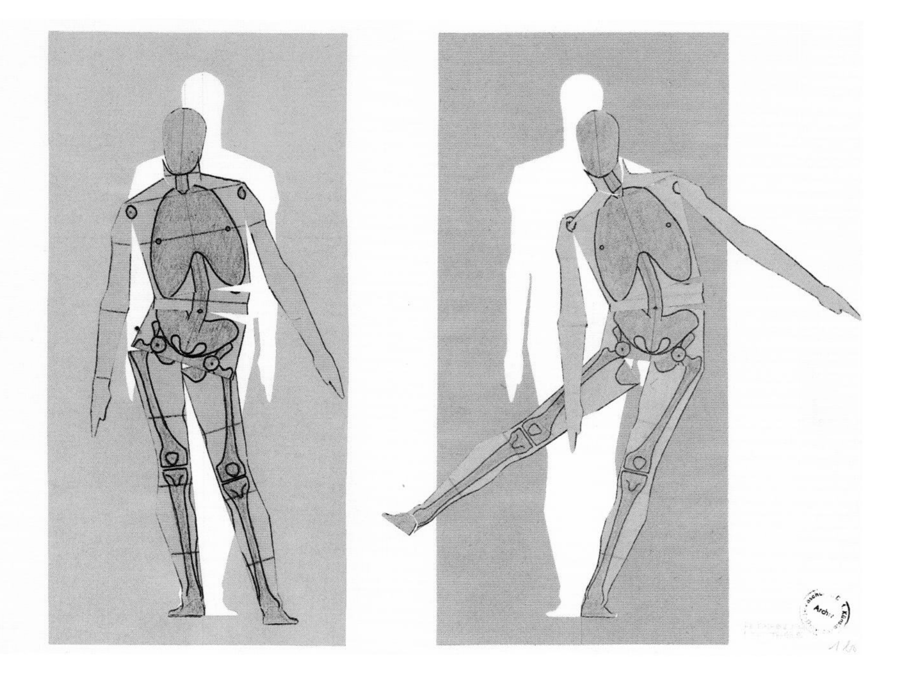

18 THE UNITY OF PROPORTION AND FUNCTION

The success of this study compared with fig. 17 is enhanced by the drawing of simple skeletal forms into the geometrized sections of the body, encouraging closer observation of the pivotal points of movement and making sense of the shape of the outer contours.
Student of painting/graphic art, first semester

19 TWO-DIMENSIONAL MODELS OF MOVABLE PROPORTIONED FIGURES IN ACTION

Besides concentration on proportional factors, an understanding of the way in which body is constructed is promoted by determining the actual pivotal points of movement. At the same time, moving positions provide information about proportional relationships and changes.
Student of set painting, first semester

20 PRINTS OF PROPORTIONED FIGURES IN MOVEMENT

Keeping to the geometrized basic forms of the body in producing the printing block makes expression of movement a playful discovery of fundamental behavior of form.
Special school of painting and graphic art

21 DRAWING MOVING POSES USING BROAD CHALK LINES

These studies are closely linked to previous investigations of proportion. It is now important to render forms concisely so as to grasp the important formal accents that develop in movement.
Student of painting/graphic art, first semester

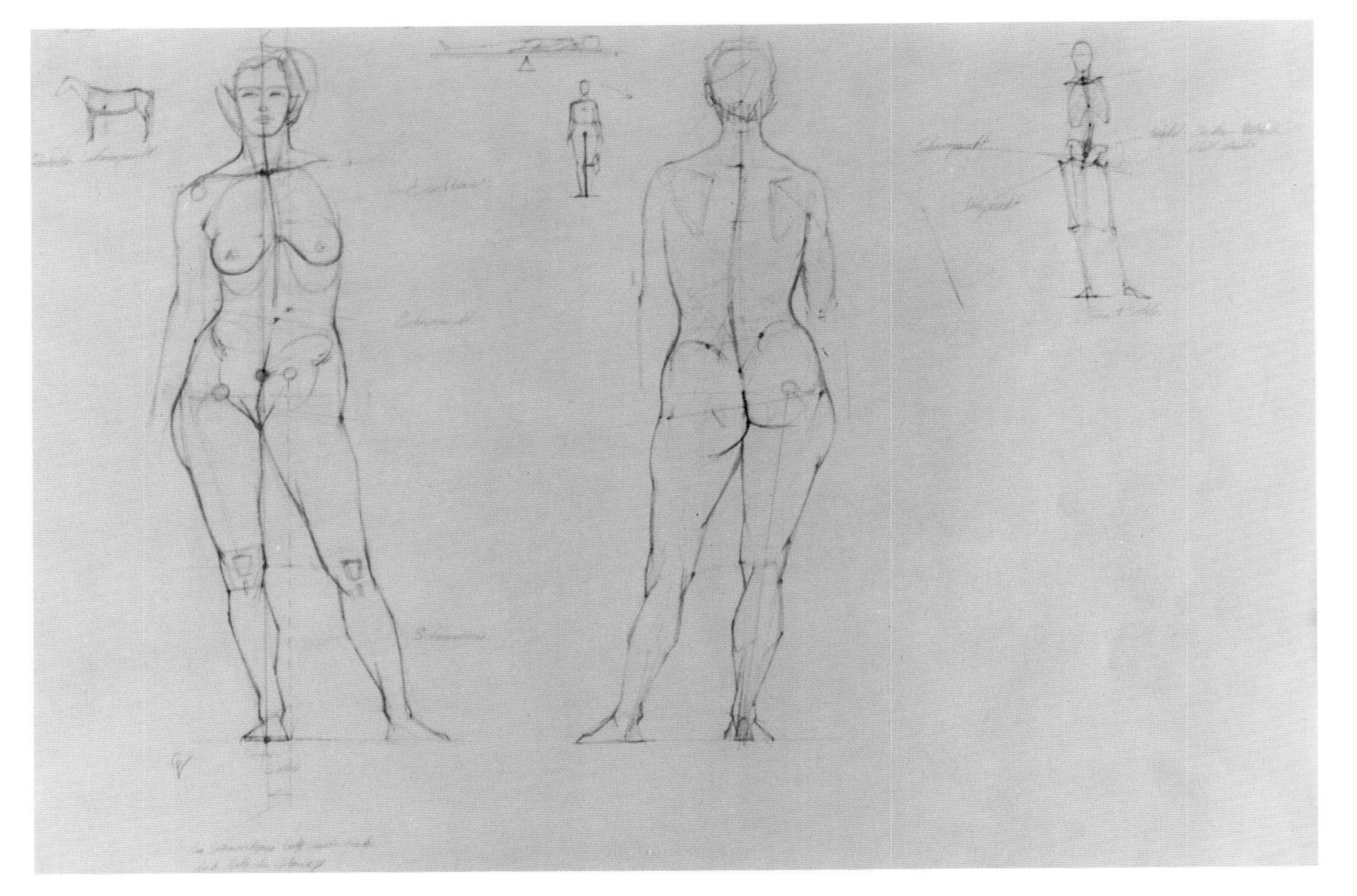

22 GRAPHIC AND LINEAR CONTRAPPOSTO CONSTRUCTION

The structural and proportional framework serves as a basis for introducing refinements of form into a larger whole. The construction starts from the weight-carrying sole and the central point of the mass (center of gravity) lying above the middle of the sole.

From the Salzburg intensive course, 1988

23 EXPRESSING MOVEMENT IN A STANDING POSE USING A PAINTBRUSH

This is a very rapid study aimed at seizing the essentials of the movement, making much use of abstraction. The directions of the limbs and body make the functional circumstances clear in an elementary way.

Demonstration study by the author, from a Bammes course at the Schule für Gestaltung, Zurich

24 STUDY OF MOVEMENT USING A GRAPHITE STICK ON ITS SIDE

One of the model's constantly changing poses is captured at lightning speed with a graphite stick used on its side, with the help of visual memory.

Demonstration study by the author, from a Bammes course at the Schule für Gestaltung, Zurich

25 EXAGGERATION OF FUNCTIONAL EXPRESSIVENESS IN STANDING POSES

Cutting out details makes it easier to discover the expression of various forms of standing, some drawn from the imagination.

Student of painting/graphic art, first semester

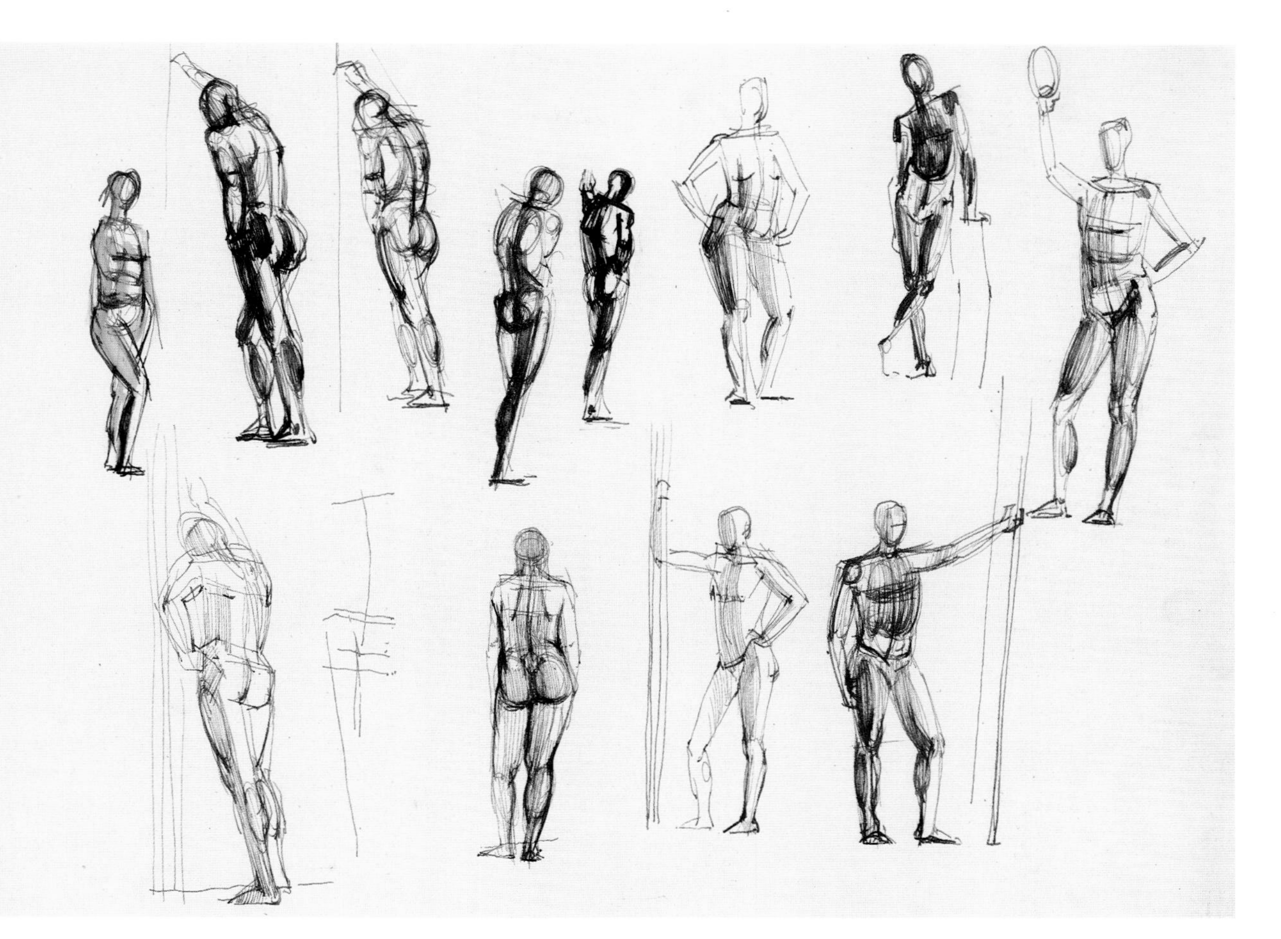

26 STUDIES OF STRUCTURE IN CONJUNCTION WITH SIMPLE THREE-DIMENSIONAL SKETCHES

In three-dimensional views of the model you must consider simply represented body volume. Here again you should avoid any clutter – learners are only too apt to get caught up in portraying a 'beautiful' nude.

Student of painting/graphic art, first semester

27 TEST STUDIES TO CHECK THE QUALITY OF SKILLS ACQUIRED

From time to time it is necessary to check your mastery of drawing skills by working from the imagination. Quite often this produces better results than short-term or prolonged observation of a model.

Student of painting/graphic art, first semester

28 SKETCH-NOTES ON BASIC FORMS OF SEATED POSES
The top row shows drawings done during discussion of proportional, structural and functional changes after a picture on the board. The bottom row contains an attempt to interpret the form of the basic figure in rectangular format.
Student of painting/graphic art, first semester

29 CHANGING SEATED POSES
This study concentrates on the psychological impression made by seated poses, at the same time clearly indicating the connection between the pelvis and the rib cage in rapidly changing poses.
Student of painting/graphic art, first semester

30 DECORATIVE FRIEZE FORMED FROM KNEELING FIGURES

This work was done after a discussion about Greek vase painting. The figures are freely invented using a movable proportioned block.

Student of set painting, first semester

31 STUDIES OF SEATED POSES USING A PAINTBRUSH

We are again concerned with fluidity of movement; also with the significance of positive and negative figures. Preparatory drawings were strictly forbidden.

Student of painting/graphic art, first semester

32 SEATED POSITION WITH A THREE-DIMENSIONAL ASPECT

In this three-quarter view using a half-dry bristle brush a concern with intersections receding depth is necessitated. Parts of the body nearest to view are deeper in tone.

From the Salzburg intensive course, 1988

33 FLUID BRUSH DRAWING

The student takes comfortable, casually relaxed sitting as subject. Continuous forms lightly and freely executed with a watercolor brush are very well suited to the task.

From the Salzburg intensive course, 1988

34 FRONT VIEW OF GESTURE IN SITTING

The preceding studies of seated poses were in profile and with clear contours. Sitting viewed frontally requires more powerful articulation. Thus linear definition is used here as well as indication of the main broad masses.

Demonstration study by the author, from a Bammes course at the Schule für Gestaltung, Zurich

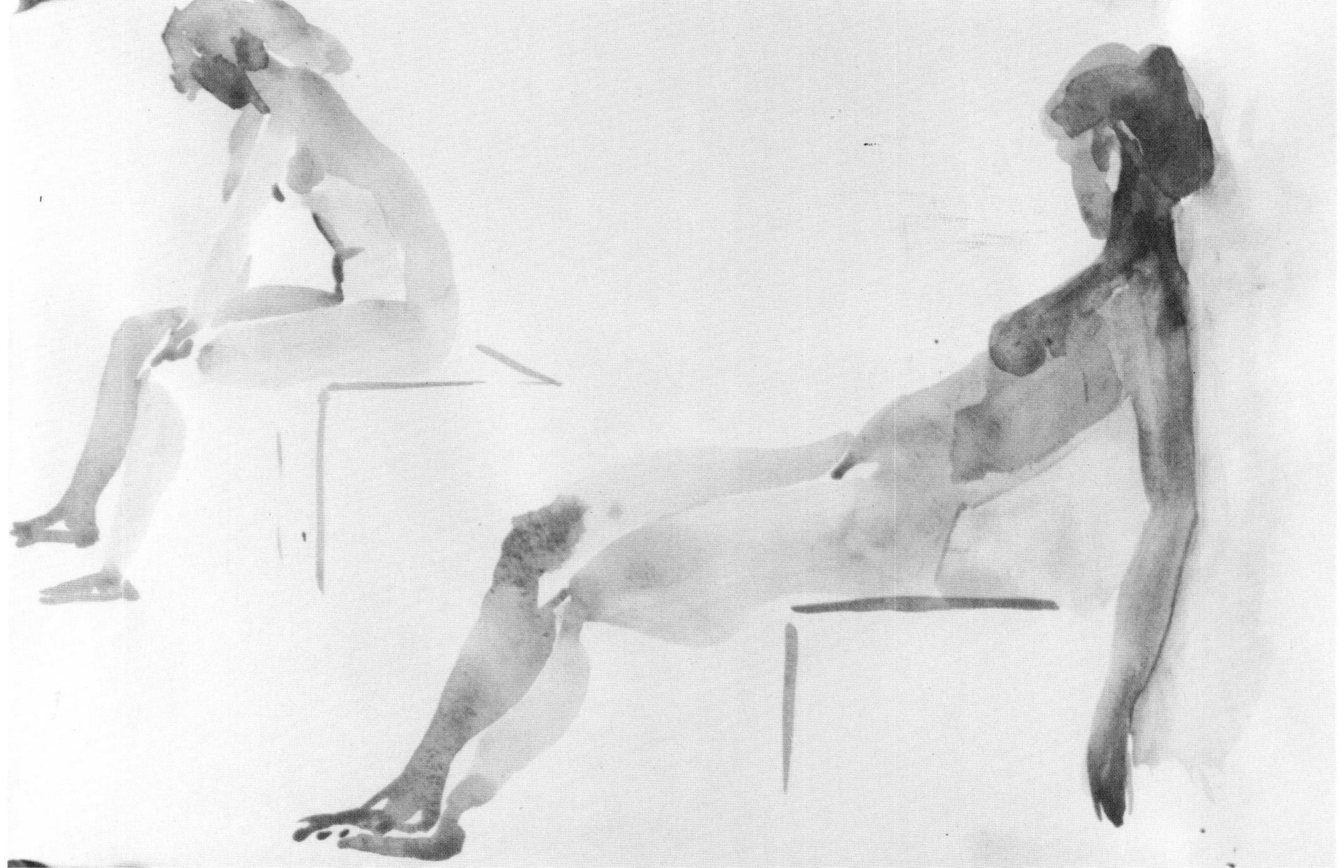

3.

Studies of locomotive and expressive movements

The transition to locomotive movements is always the consequence of disturbed balance, with the body's center of gravity shifted forward over the tipping edge (the front of the instep). A walking or running stride is essentially the rhythmic repetition of the body catching itself as it begins to fall forward.

This causes a problem for artistic work that is seldom resolved satisfactorily: the suggestion of a possible moment of fall with the body leaning forward, perhaps still at the very moment when the swinging leg is not yet ready to take the body's weight. This makes the observer feel uncomfortable. We perceive any such representation as a disturbing, frozen, chance snapshot, searching for stability. Art has a significant role to play here. For instance, if we were mentally to convert a monumental painting such as Hodler's Departure of the Jena volunteers, with the double support of the wide apart legs, into the instantaneous situation of falling, what a disaster it would be! We accept lively, illustrative graphic art far more readily, of the type used in Slevogt's Leatherstocking illustrations, for example. Italian Futurists opted to express relative movement or virtual forward movement through phases of movement represented simultaneously, which go through the body and seem to lift its physical weight.

Strictly speaking, most representations of a stride are only illusionistic movements, irrespective of whether only one leg is functioning as a supporting, bracing member and the other as a swinging leg, or both legs are touching the ground at the same time. You need only call on your own kinesthetic sense to know whether you could not also hold still in such a position. Usually the answer is yes.

The behavior not just of the legs (figs 35–37) but of the upper body too – its upright (figs 36, 37), forward or backward position reduces or reinforces the appearance of motion, the upright position suggesting a brisk march, the forward position haste, and the backward position with the body behind the front foot, dawdling or hesitation.

Both figs 36 and 37 confirm that, in accordance with the decorative character of the printed work, the specific tendency in the student's future profession toward enhancing real dynamics of the stride, to the extent of exaggerating its length (fig. 37), means that these are not genuine movements. The dancing movement (fig. 35) shows the transition toward expressive movement. Essentially it expresses an atmospheric situation through the simultaneous movement of the limbs for no explicit purpose.

These three decorative works, produced after looking at an antique vase painting, demonstrate a playful use of printing with movable proportioned blocks, but are not so well adapted to conveying the course of real movement. The reverse is true of figs 38–41. It is obvious from fig. 38 that both the dynamic and psychological components of walking, running and gesture had previously been discussed. Even the relatively small figures quickly drawn with chalk and brush, especially in their phased sequences, do not call the 'reality' of these instantaneous representations into question, even those of falling movements.

To a large extent the studies illustrated in figs 38–39 are based on visual memory, visualization and sympathetic insight into movement covering hesitant, hasty, dragging, dancing, tired, happy, vigorous or affected strides. It need hardly be pointed out that the student can build on particular discussions of structural and dynamic rudiments as criteria for assessing whether figures are standing or falling, stable or unstable. These criteria are also applied when undertaking lively imaginative work.

The situation is similar in the studies illustrated in figs 39 and 41. The model who collaborated sensitively in the work was constantly moving so that students had no time to go into detail.

The model pulled, lifted, pushed, fell over, carried weights above his head or her back, sometimes standing still and sometimes moving forward. A suitable medium for conveying this is the brush which can react quickly. It should be repeated that the round, well filled watercolor brush works quite differently from the pencil (fig. 40). When using a brush you need only apply it and you have a volume, draw it along and you have a complete arm, let it trail and the volume tapers. However, if the brush is constantly lifted off the paper and reapplied, no flow of movement will be created and you will have only splodges. The esthetic appeal will have been lost. In setting out to convey form and expression the pencil study (fig. 40) can work with whole bundles of lines, or if the person drawing is confident he or she can define the flow of movement with a single line. The tip of the pencil is also useful when you need to go into individual aspects in more detail or to clarify a three-dimensional effect, foreshortening or an intersection. If you already have some experience of studying the skeleton, you can use the study of movement analytically to examine elementary aspects of function more thoroughly.

In the chapters on proportion, structure and movement we have shown the possibilities of a combined figural effect and the figural priorities that you should aim for. In future studies these are generally not mentioned expressly. They must be assimilated into your work 'silently', present and available as a constant means of self-correction.

35 DECORATIVE PLAY WITH EXPRESSIVE MOVEMENTS

Printing with proportioned blocks representing the moving parts of the body stimulates an inventive approach to movement and needs to be activated by our kinesthetic sense.

Student of set painting, first semester

36 STRIDING POSTURE SUGGESTING A PEACEFUL MARCH

The upright posture of the upper body, the bent front leg placed supportingly on the ground and the offloading back leg combine to convey a relaxed stride. Printing with movable proportioned blocks is used as in fig. 37.

Student of set painting, first semester

37 EXAGGERATION OF THE STRIDING POSTURE

The wide spacing of the legs makes the movement appear extremely brisk, but in actual fact it would not allow the rear leg to swing over past the dead center.

Student of set painting, first semester

38 PHASES OF THE STRIDE WITH DIFFERENTIATED MOVEMENTS
It is not only the movement of the legs that provides information suggesting speed or hesitation; the position and form of the upper body also come into play.
Student of stage design, first semester

39 LOCOMOTIVE MOVEMENTS
The directions used in individual activities are also of fundamental importance here in expressing the movement the student wishes to convey. But the flow of the brush is too often disrupted in making contrary series of lines.
Student of sculpture, first semester

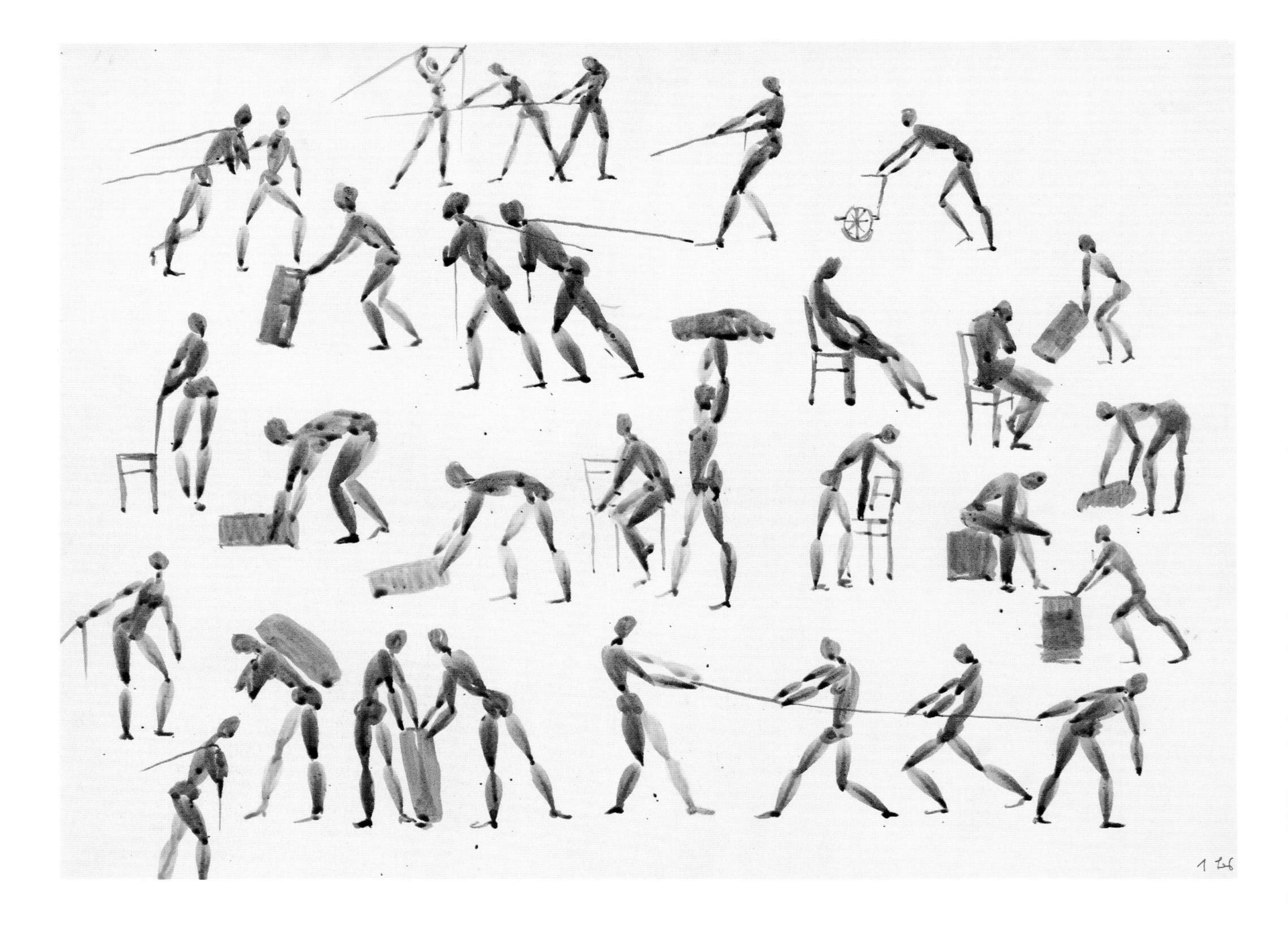

40 WORKING MOVEMENTS EXPRESSIVELY EXAGGERATED

The fact that a heavy load is being pulled is made clear by the way the body leans far forward and the bracing leg is stretched out far behind. Käthe Kollwitz's Plowmen was used as a source of expressive inspiration.

Student of painting/graphic art, first semester

41 RECORDING MOVEMENTS WITH THE BRUSH

This work is more fluent than fig. 39, concentrating mainly on the interconnected expression of movement in rapidly changing actions by the model.

Student of painting/graphic art, first semester

4.

Studies of the skull and head

Whereas the three previous chapters have been concerned with the figure as a whole in terms of proportion, construction and motion, we now come to the structural and anatomical studies of the head as a module of the whole figure.

Once again, in studies of the skull and head, establishing sizes and the connection of the two sections of the skull – the cranium and face – are of prime importance. The basic forms of these volumes are depicted with the axis of the eyes halfway down the head height (crown of head to point of chin, fig. 42). Only after becoming thoroughly conversant with the functional and structural systems of the skull in profile and front view should the student embark on a study of the skull (figs 43–46). For this you should draw up a three-dimensional reference system (based on the central vertical axis and subsidiary cross axes). Using this three-dimensional framework for conveying the three dimensions, construct a view of the basic volumes of the cranial box and the facial skeleton two-dimensionally; only once this has been done should you move on to subtle, detailed differentiations.

The fundamental principle is always to create unambiguous planes to view, establishing the front, side and 'roof' planes in their relative positions. They reflect the groundplan of the base of the skull depicted in perspective above which the dome of the cranium is situated. All concealed edges and corners should be drawn in as intersections (fig. 45).

The skull is a marvelous piece of architecture and its facets should be observed and drawn with great attention. It is a building with supporting columns, consoles, recesses etc. Only by drawing the skull constructionally will you really discover the foundations on which the living appearance rests.

After the shapes of the mouth, nose, eye and ear have been considered, individual studies of them should be drawn (figs. 46, 47), and finally studies of the head as a whole (figs. 48–52). In undertaking the individual studies and in drawing the head strict attention should be paid to structure (fig. 46):

- The nose has a front ridge, side planes, nostrils and a tip which are joined to one another by facets (be careful with the intersections at the nostrils).
- The visible part of the eyeball is curved as part of a sphere with bulky covering parts (the upper and lower eyelid). Here again you should carefully observe the way in which the eyeball and the lids overlap (figs 46, 47).
- The mouth is a three-dimensional, curved shape, especially in the notched area of the red of the lips, and protrudes and recedes (figs 48, 50, 51). In three-dimensional views (fig. 51) both the overlaps within the red of the lip and its foreshortening, and the intersections with the cheeks must be mapped out.
- The ear is not just a flap hanging down somewhere or other: it is a folded, thin cartilaginous substance with the contours of a curled-up spiral which is an articulated, shell-shaped, three-dimensional space (so a special study is called for!).

We begin studies of the head, ideally using your own head with the help of a mirror, by establishing proportions (figs 48, 49) of the position of the axis of the eyes (the center), the length of the nose, forehead and lower face; each is usually a third of the length of the face measured from the chin to the hair line. Once these measurements have been established, check the widest measuement of the face and skull (usually $^2/_3$ H). Before proceeding any further it is a good idea to grasp these basic proportional factors really well, both in principle and two-dimensionally (fig. 49), and you are then ready to move on to modeling (figs 50, 51). The experience and skill you have gained in drawing the skull are applied again here. Work out the spatial gradients of the planes and how they meet corners and edges: it is better to make the junctions relatively hard-edged rather than too indeterminate. The volumes are encased in convex planes, and these surfaces meet one another. These are events involving form, meetings between forms. There are no intervening concave dips between the convex shapes leading from one to another: they collide. If we ignore this, the head becomes a wishy-washy lump of soap. What is called for here is articulation of form.

Expanding on the drawing of the head (which should not be a portrait study) again provides scope for the power of visualization and imagination, first with different physiognomies in profile (figs 53, 54) then proceeding to striking head types (fig. 55) and grotesque heads from fables, fairy-tales or the theater (fig. 56).

The physiognomies based on visualization and imagination should not be pure fantasy with no plan or program. Start with a striking profile view (fig. 53) and, like Dürer, draw the falling facial line, i.e. the line running from the highest point of the forehead down to the incisor teeth, which establishes the slant or tapering of the profile (fig. 53, top left). This factor alone opens up an infinite number of physiognomic possibilities, if you also shift the relationships of the elements of the face and include the full range of variation in nose and mouth shapes.

In three-dimensional representations of the head, the imagination can be stimulated and guided if you try to achieve the maximum consistency in portraying the typical features of a slim, lean person, an athletic, robust one or someone who is squat and rotund (fig. 54). Otherwise you will end up producing some weird characters.

Even when drawing grotesque heads, and Leonardo da Vinci provides some fine examples, a proper measure of anatomical possibility should be retained in all deviations from the norm (fig. 56).

Using your imagination and powers of visualization when drawing and designing the shapes of the head is important in order to preserve the form and appearance of the head and body.

However, beginners should not be seduced into trying to turn their first drawing of the head into a portrait. The head should be thoroughly understood graphically before an individual personality is elaborated in a portrait.

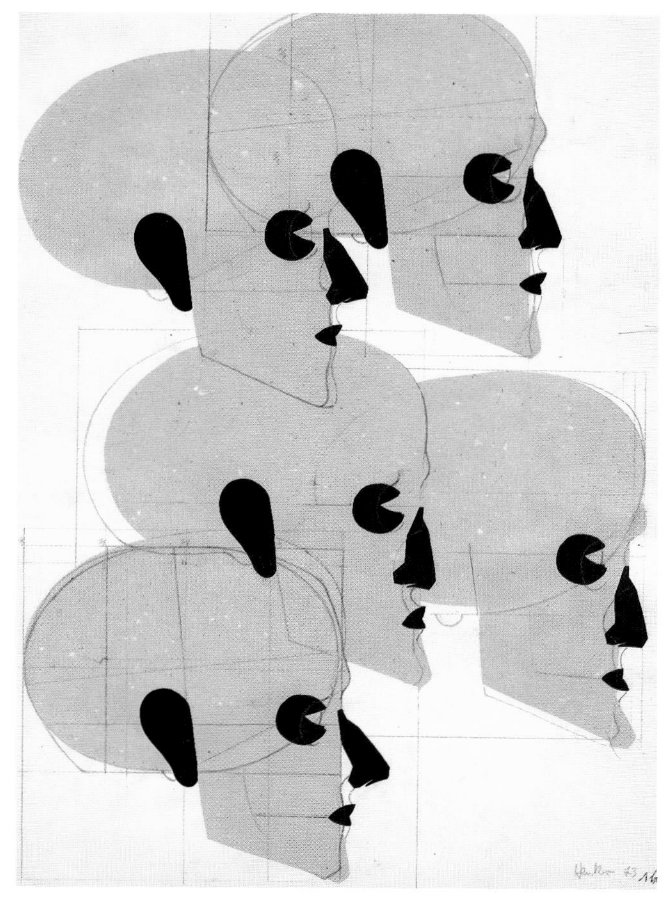

42 STUDIES OF THE PROPORTIONS OF SECTIONS OF THE SKULL
Using collage you can link the ovoid cranial shape with the vertical shape of the facial skeleton, incorporating a physiognomic effect and nose shape protruding to a greater or lesser degree from the profile of the face.
Student teacher, first semester

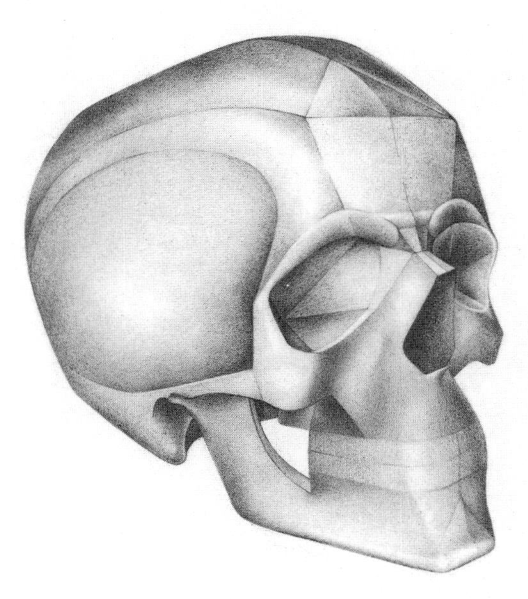

43 ARCHITECTONICALLY CONCEIVED THREE-DIMENSIONAL STUDY OF THE SKULL
It is the structure of the basic parts and the constructional and functional assembly of the individual components (e.g. the jawbone, nasal tunnel, the column running from chin to forehead) that lead to successful execution of the task, not drawing the skeletal details.
Student of restoration work, first semester

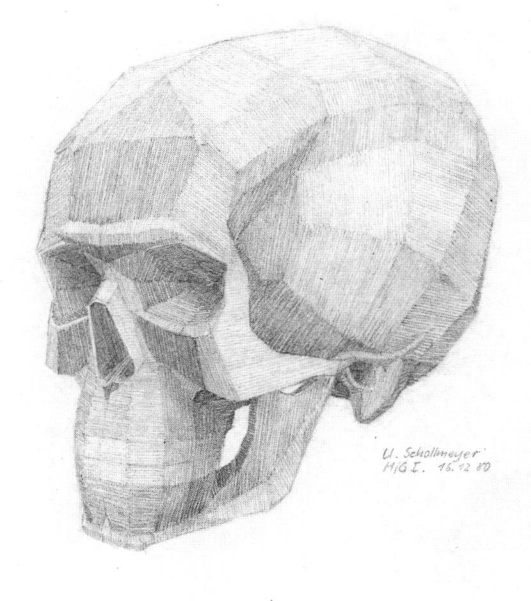

44 USE OF GRAPHIC EFFECTS IN ARCHITECTONIC DRAWING
The sloping planes enclosing the skull are drawn in with a network of lines running in skillfully aligned directions with maximum contrast to reinforce the solidity and three-dimensionality of the skull.
Student of painting/graphic art, first semester

45 GRAPHIC PENETRATION OF THE SKULL SHAPE

Starting from the three-dimensional reference system, draw the simplified forms of the base of the skull and the skull cap resting on it, then the middle and side columns of the facial skeleton. You are thus compelled to draw through all overlaps. In purely linear studies of this kind carefully observed intersections are particularly important.
Student of restoration work, first semester

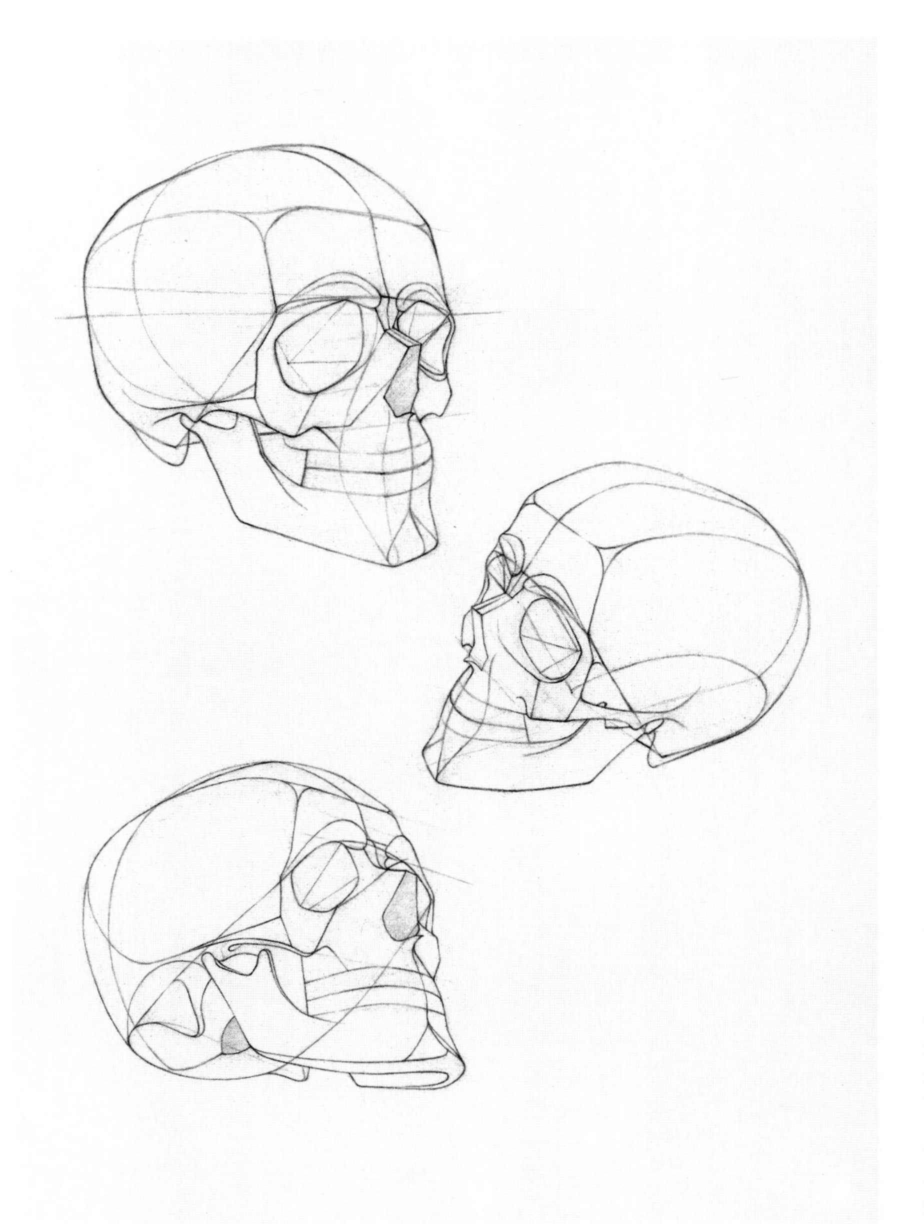

46 STRUCTURAL INVESTIGATIONS OF THE SHAPES OF PARTS OF THE HEAD

As in drawing the skull, the forms of the soft, fleshy parts are also conveyed by the main planes and gradients on view, building on experience gained in structural drawing of the skull.
Student of painting/graphic art, first semester

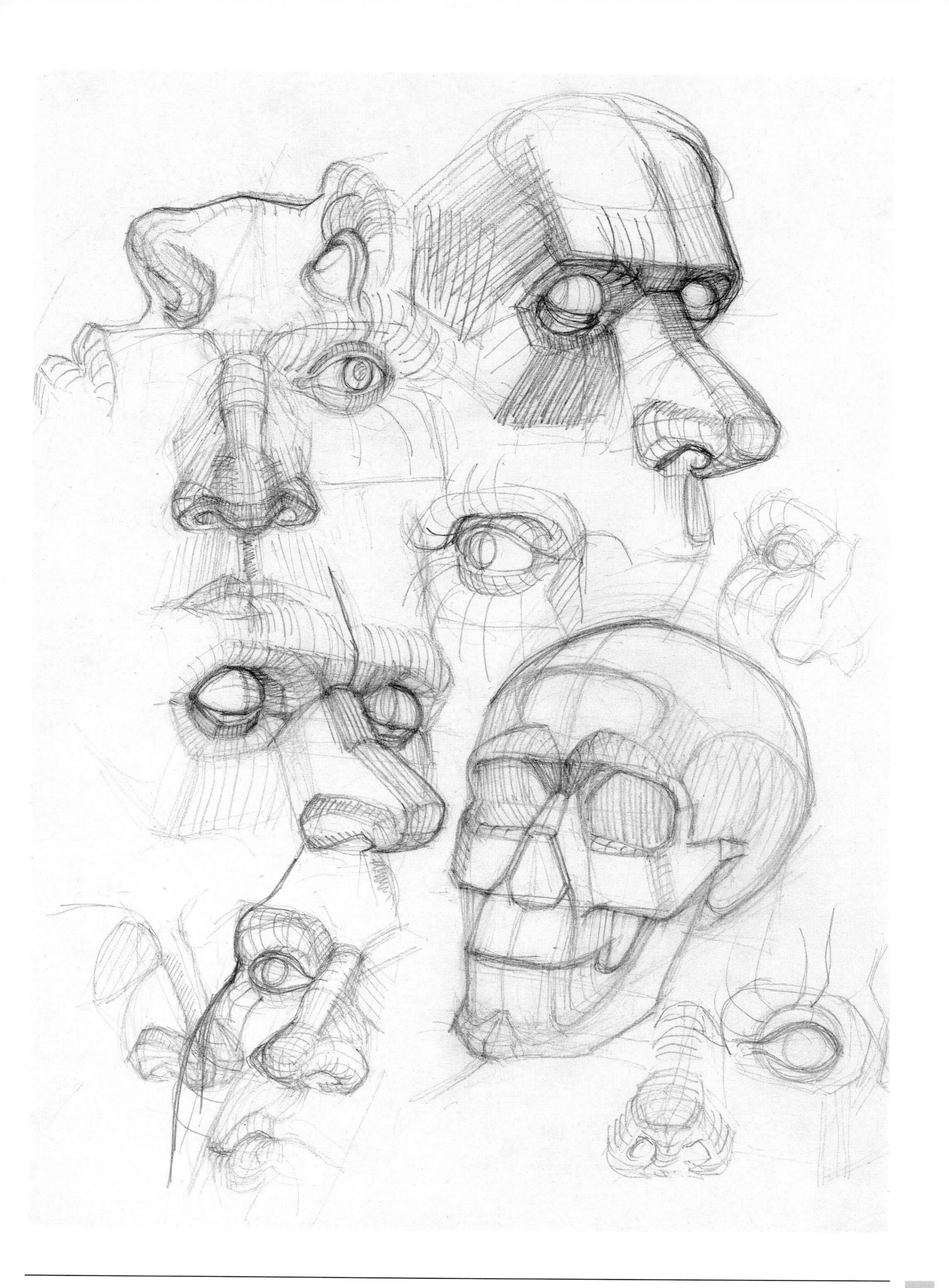

47 STUDIES OF THE MODELING OF THE EYE AND THE SURROUNDING AREA

The assignment involves understanding the modeling of the eye as part of the surface of a sphere which is imbedded in the space enclosed by the bridge of the brow, the side of the bridge of the nose and the jawbone.

Student of stage design, first semester

48 RECONSTRUCTION OF THE SHAPE OF THE SKULL IN A SELF-PORTRAIT BASED ON VISUALIZATION

Within the contours of the individual head form, checked by measurements of proportions, the shape of the skull underneath can be found. Yet another indication of the extent to which the form of the head is determined by the basic skull.

Student of restoration work, first semester

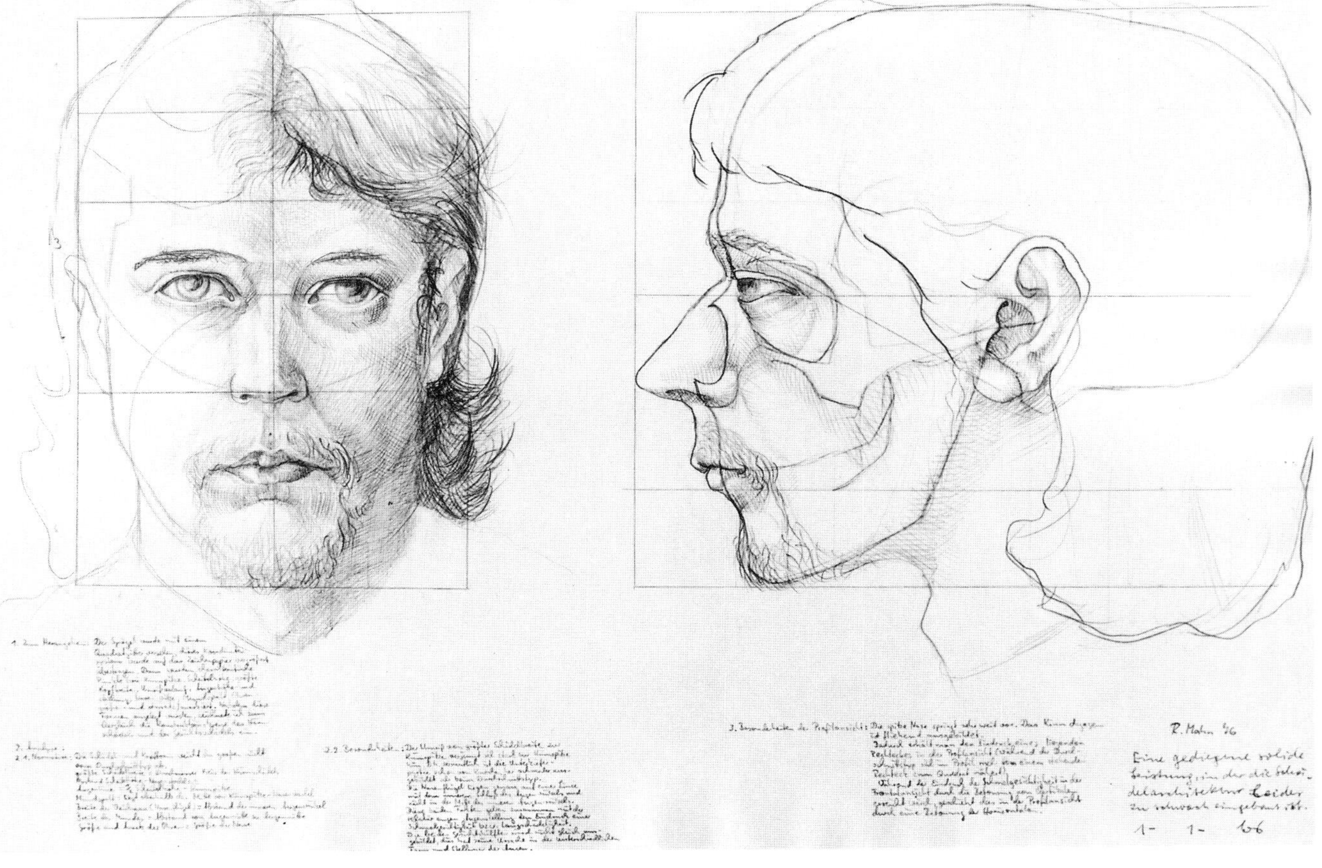

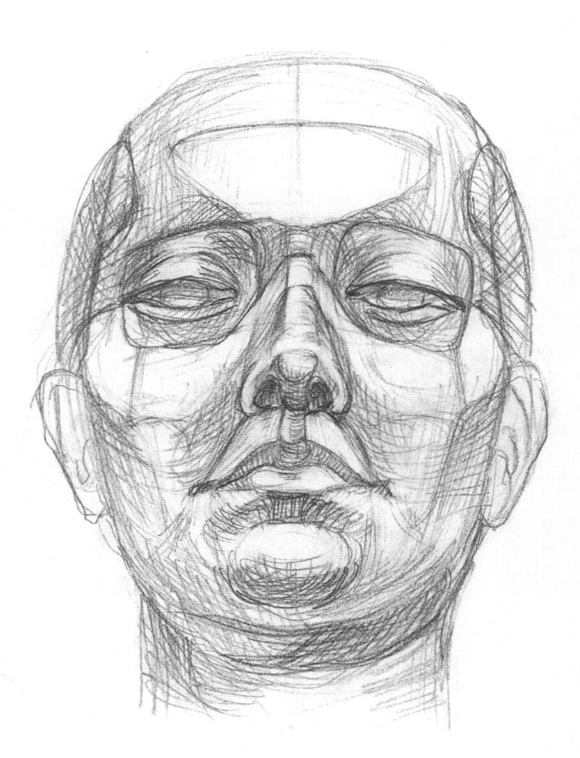

49 PROPORTIONING THE HEAD AS THE BASIS FOR INDIVIDUAL FEATURES OF THE FACE

In drawing, the unique appearance ('likeness') of an individual can be created simply from careful measurement of the lengths and breadths of the face, and the position, shape and size of the component forms, without making any further attempt at differentiation.

Amateur artist

50 CONSTRUCTION OF THE PLASTIC APPEARANCE OF THE HEAD

If establishing the proportions (see fig. 49) is linked with sketching in skull forms and their facets, the volumes of the individual forms can be fitted into the formal solidity of the whole head in a completely stable way.

Student of restoration work, first semester

51 DEVELOPING THE MODELING OF THE HEAD FROM THE STRUCTURAL SHAPES OF THE SKULL

Here again the 'self-portrait' is used to gain quiet self-confidence in a general, basic drawing of the head; the face is treated as an assembly of the forms of the parts into an integral whole.

Student of painting/graphic art, first semester

52 ASSIMILATING THE STUDY OF THE HEAD
With side lighting constituent forms are brought together to make larger parts; gouache has been used on a wet ground.
Special school of painting and graphic art, first semester

53 ATTEMPTS AT INVENTING PHYSIOGNOMIES
The course of the profile of the face is determined to a large extent by the basic direction of the face line (forehead-lip) which indicates a steep or a flatter profile. Within each profile the sections of the face have been varied in length.
Student of stage design, first semester

54 IMAGINARY PHYSICAL TYPES
Discussions on the characteristics of different types of physical build had taken place, which are reflected in the drawings in a free treatment of the face as a unit.
Student of stage design, first semester

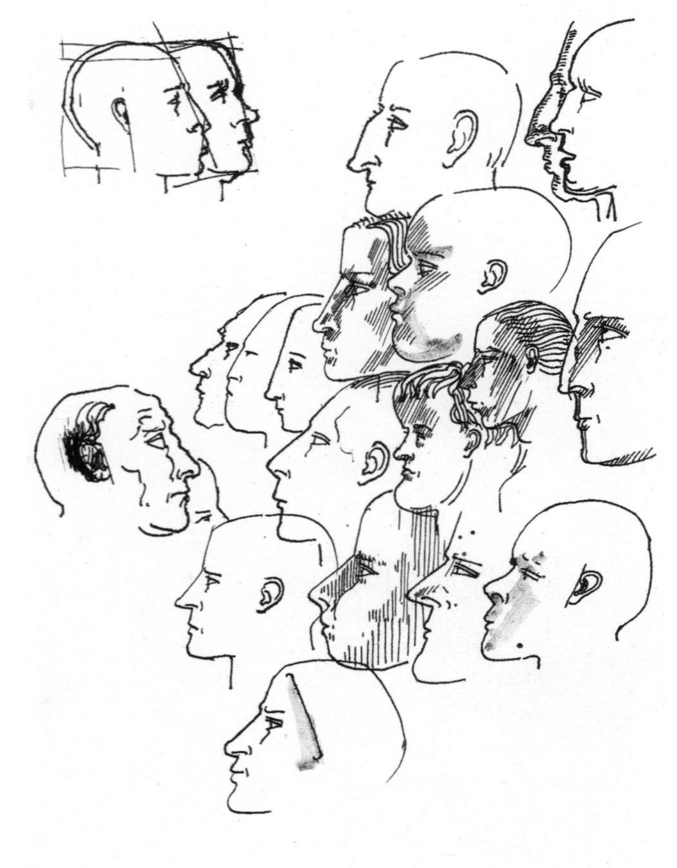

55 STUDIES OF AN ACTUAL PHYSICAL TYPE

The individual characteristics of the Mongolian race are established in depicting a person with Mongolian features.

Student of stage design, first semester

56 IMAGINARY GROTESQUE HEADS

Intrinsic to the light-hearted assignment of sketching figures was the maxim that features, despite their abnormality, must still be organic shapes.

Student of stage design, first semester

5.

Studies of the construction, function and plastic behavior of the leg

Before hoping to become proficient in conveying the moving mechanics of the body in drawing, in this case those of the leg, the student should undertake a careful review of the overall construction of the body, the forms of the joints and their mobile muscle forces. All skeletal shapes should be simplified so that the constructional form can tell us something about function:

- The most powerful joint is the knee joint which works on a hinging principle, with a roller (a double roller with a projecting front surface for the knee-cap) which can roll over an incongruent corresponding form, the top of the shinbone or tibia, causing the joint to open (fig. 57). Constructional simplifications of this kind help us to recognize what is happening.
- The constructional simplification of the pelvis is based on the container-like character of the lesser and greater pelvis (forming a funnel shape), with the front part dropping down gradually. This creates the plastic reference points of the pubic bone and the two front spines of the hip bone (plasticity of the stomach). The crest of the hip bone then runs out sideways and on round to the back. The pelvis is a constant spatial area, and its sexually specific shape forms the body's center of movement (figs 58–60).

The pelvis, thigh bone or femur and knee-joint considered as one overall unit form the functional basis where the muscles of the thigh-bone start and are attached. After considering these, we must examine the basic form of the live knee (figs 61, 62) in detailed studies, especially the knee skeleton's capacity for changing in motion and the groups of functional muscles between the pelvis and the shin bone. We shall examine the planes to view and the three-dimensional distortions caused by all factors affecting form. Bearing cross-sections in mind makes it easier to organize the hatching used for modeling. Working out the structural interplay between firm skeletal forms and soft, fleshy forms should prevent the knee from turning into a shapeless mass.

Once the skeleton of the leg as a whole and its musculature have been learned, with full studies of the skeleton of the foot (fig. 66) and the thigh bone (figs 64, 65), the student has sufficient knowledge and preparation to be tested in a three-part intermediate exam partly based on visualization (figs 63, 67, 68). An understanding of the construction of the skeleton of the foot (fig. 66) with its connecting, jointed forms, its character as a recessed vault with cross-vaulting and lengthwise vaulting, is just as important for the structure of the living foot as the skull is for the head. If the construction of the skeletal vaulting of the foot is ignored, a drawing of it can all too easily look like a shapeless slipper.

Finally come studies of the leg as a structured living form at rest and in motion (figs 67–76). The examples show a variety of ways in which it can be represented, but the following criteria are valid for all of them:

- Clarification of the viewing perspective (eye level = horizon level) so that perspective alignments can be based on it (views from above or below).
- Clarification of the space enclosed between the two legs.
- Clarification of the form of the various basic volumes (e.g. conical and cylindrical forms) to ensure that they connect correctly.
- Elaboration of the principal forms so that subsidiary forms can be integrated into them.
- Deduction as to how the forms move (short, abrupt convexity as against gentle, extended convexity – taut forms as against curved ones).
- Use of anatomical knowledge to pinpoint the precise location of the spaces – depressions, furrows, grooves, cavities (e.g. the hollow of the knee when bent) – powerfully or gently impressed into the volumes (see especially figs 71, 72, 73, 76).
- Not least, elaboration of the functional expression: loading or releasing of a load, supporting and supported, bending and stretching, and how the joints behave plastically. What is the effect of loading, pressure (e.g. fig. 75) and stretching on the soft, fleshy forms?
- Finally, working out the sexually specific formal character of a male and female leg.

It is up to each individual to meet all these criteria in his or her own way. Each student is free to decide whether to work with constantly irregular hatching, hatching that models the body, with cross-sections only, purely with line or with regularly applied parallel hatching, and whether just to sketch or to do a finished drawing.

Close observations of the stucture of the basic forms of anatomical features – the skeleton, the pelvis, knee, foot and their joints, muscle systems – from life, and conversely working from life back to an anatomical analysis, are indispensable. Graphic insight into essence can be achieved only once it has really been understood. Otherwise it is all too easy to remain at a superficial level.

In all our previous graphic studies, rather than being content to work only from a model who was actually present, we extended our field of action to drawing based on visualization. This should be done with the leg too, and you should also loosen up again with an imaginative approach to the form (fig. 78). A thing can only be considered to be understood if you can draw it from the visual imagination. When students can work successfully from the imagination (fig. 77), this shows that they have understood a form as difficult and changeable as the knee-joint in its various functions and contractions.

57 CONSTRUCTIONAL FORMS OF THE SKELETON OF THE KNEE IN ACTION

Simply copying the natural form without understanding it is of no value in the learning process. Only by simplifying the form constructionally is one forced to look thoroughly and form permanent visual concepts.

Student of painting/graphic art, first semester

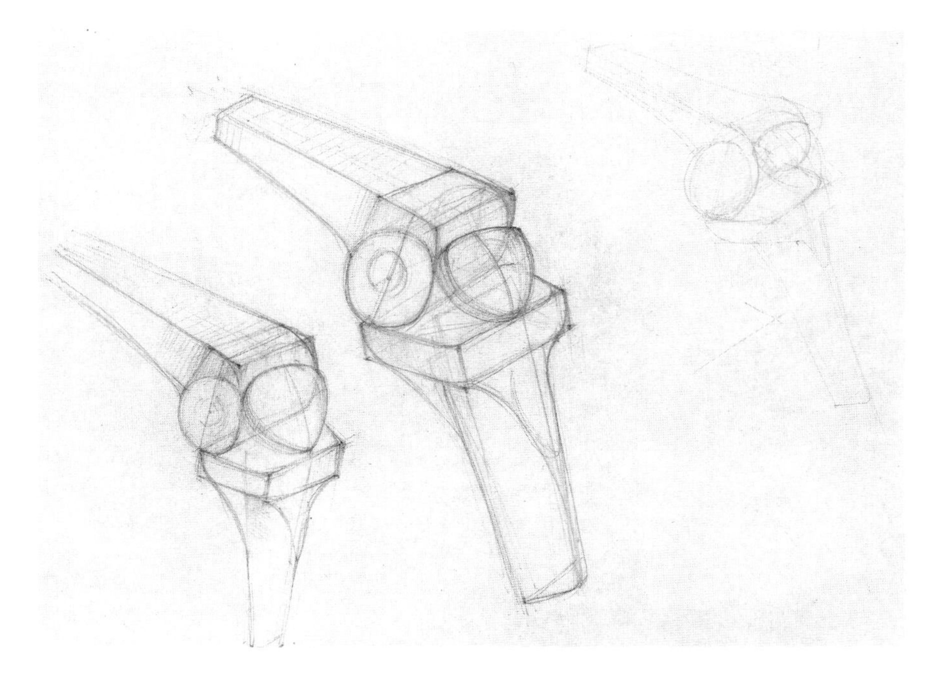

58 CONSTRUCTIONAL REPRESENTATION OF THE PELVIS AS A CONTAINER FORM

The way the torso finishes at its lower end with the pelvis is impressively reflected by many accents in the modeling of the living figure. Above all, the hollowed-out form of the front of the greater pelvis must be expressed as a basis for the spatial ins and outs of the front and side abdominal wall.

Student of stage design, first semester

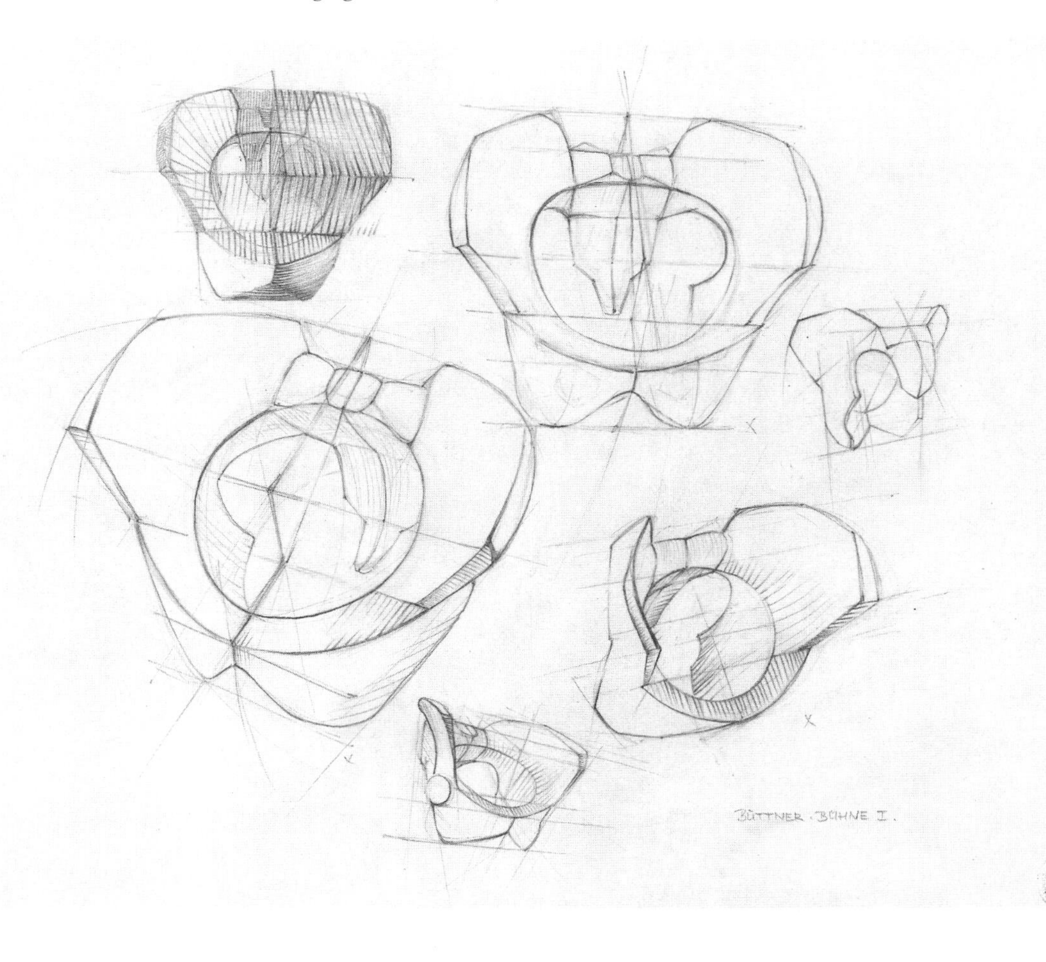

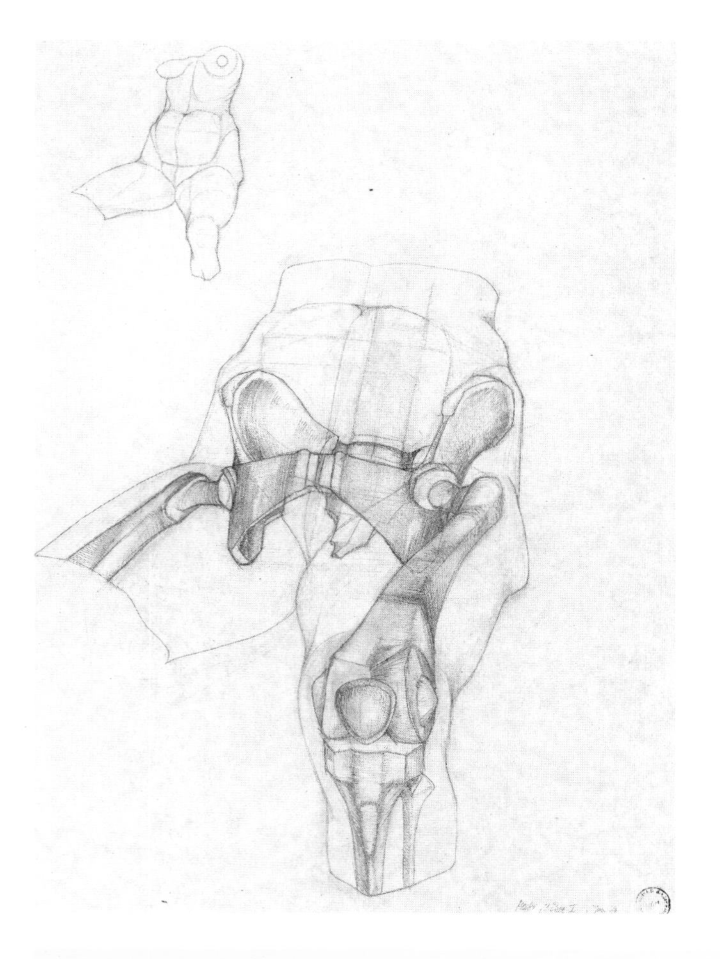

59 CONNECTION BETWEEN THE SKELETONS OF THE PELVIS AND THE KNEE

The backward-leaning pose of the nude model is analyzed on the basis of the skeleton so that the back-tipped position of the pelvis can be understood in functional terms. Thus analysis of the living form demands an imaginative, and to some extent contingent, reconstruction of the skeletal forms.
Student of stage design, first semester

60 FROM SKELETON TO LIVING FORM

Progression in relation to fig. 59 consists in the development of a pure visualization of the skeleton which can be used as an aid to finding the external contours of an assumed nude model. The spatial positions of the skeletal forms in relation to one another are also relevant.
Student of sculpture, second semester

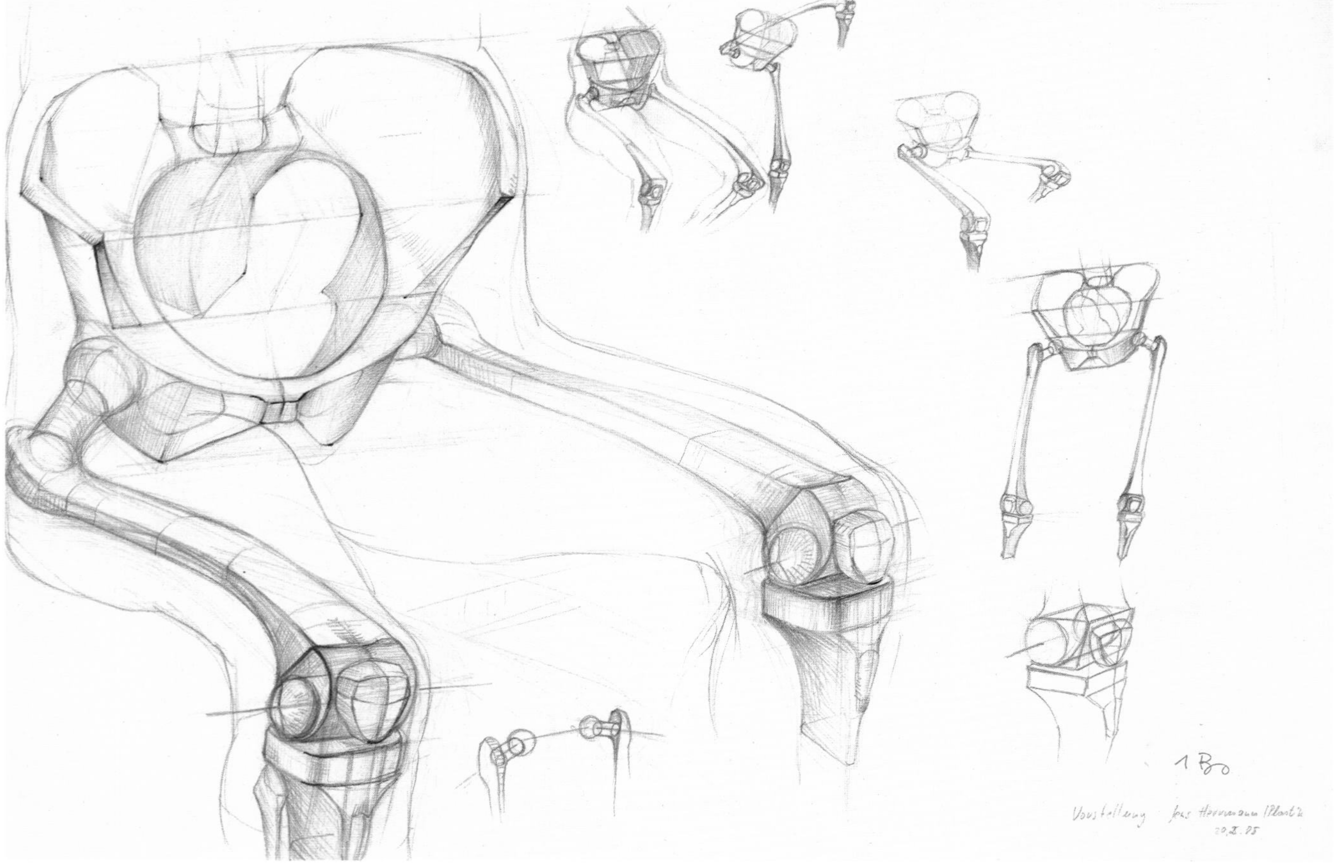

61 THE KNEE JOINT IN ITS MUSCULAR SETTING AND AS A LIVING FORM

Studies of the knee usually fail because of a lack of exact knowledge of anatomical details and how to interpret them, but also because the formal components lack clear definition. The best results are achieved by logically assessing the various planes to view. Student of painting/graphic art, second semester

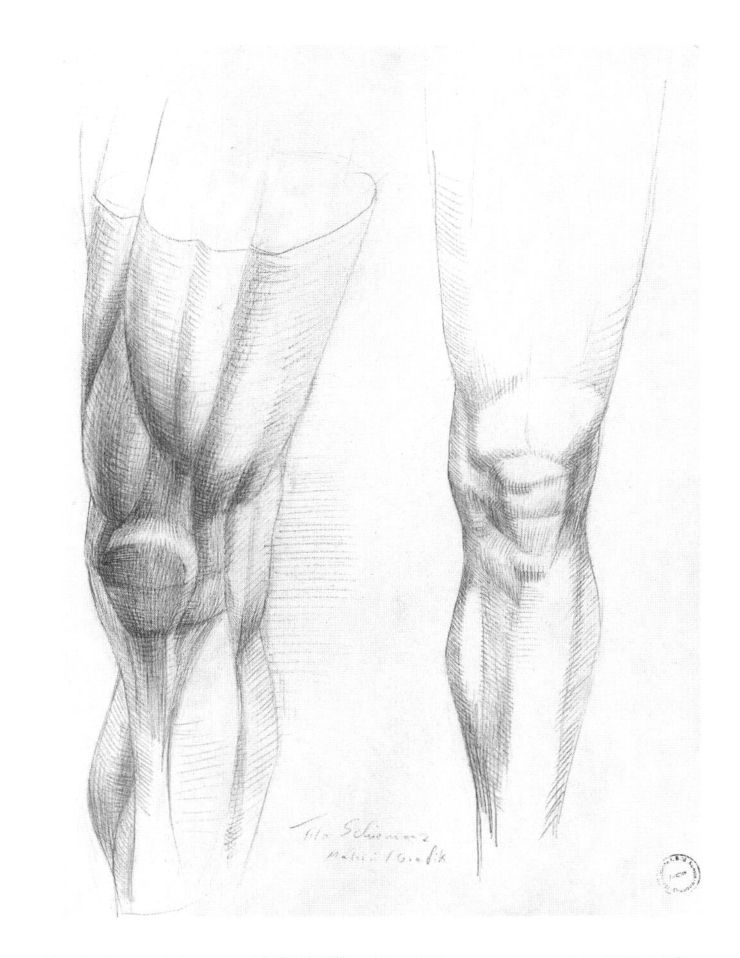

62 THE KNEE JOINT IN ITS PLASTIC CONTEXT

Embedding the joint in the outstretched leg and the sharp-edged projection of the knee-cap in a bent position are among the focal points of life studies, along with depicting the plastic distortions of the upper and lower leg by examining cross sections.
Student of sculpture, second semester

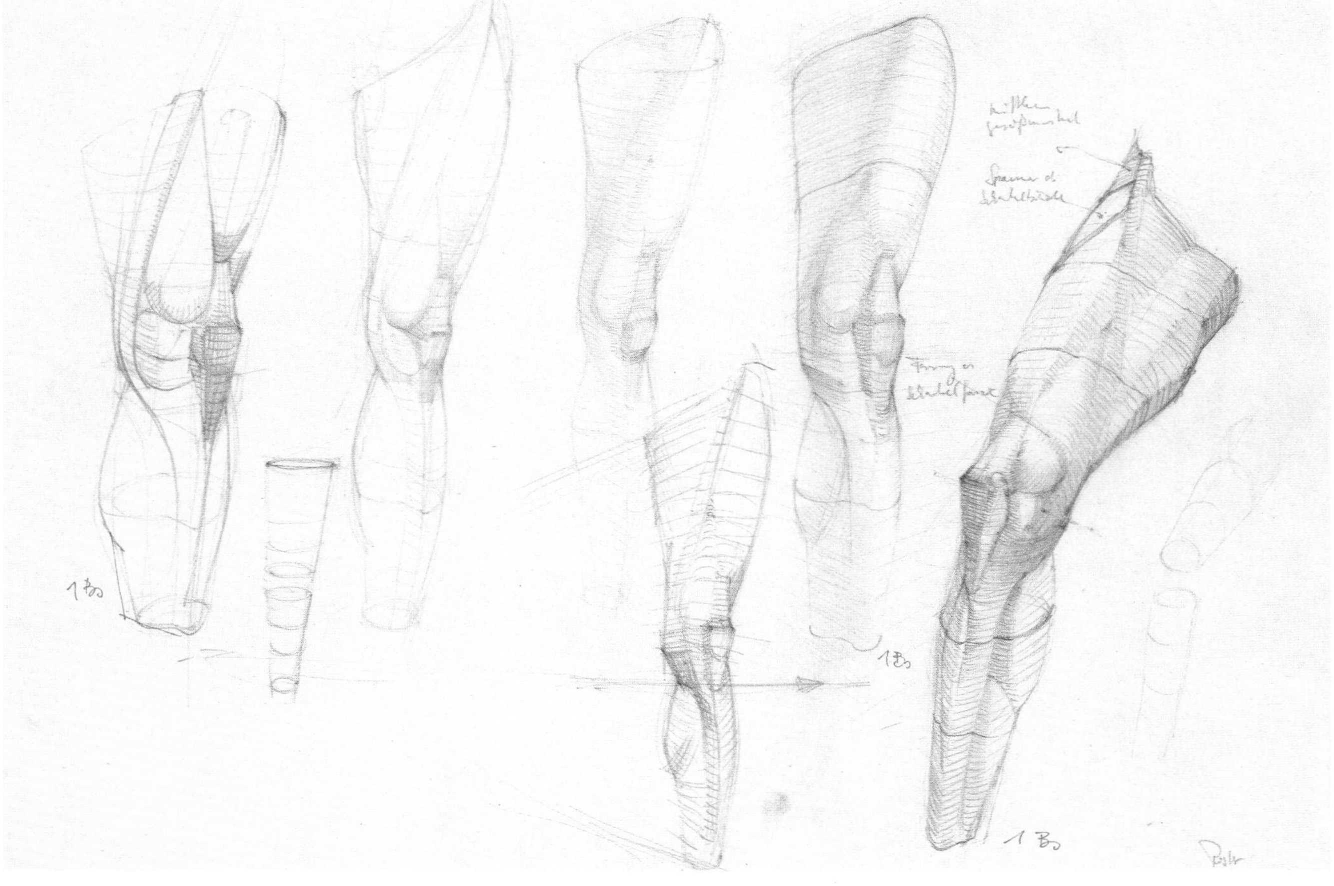

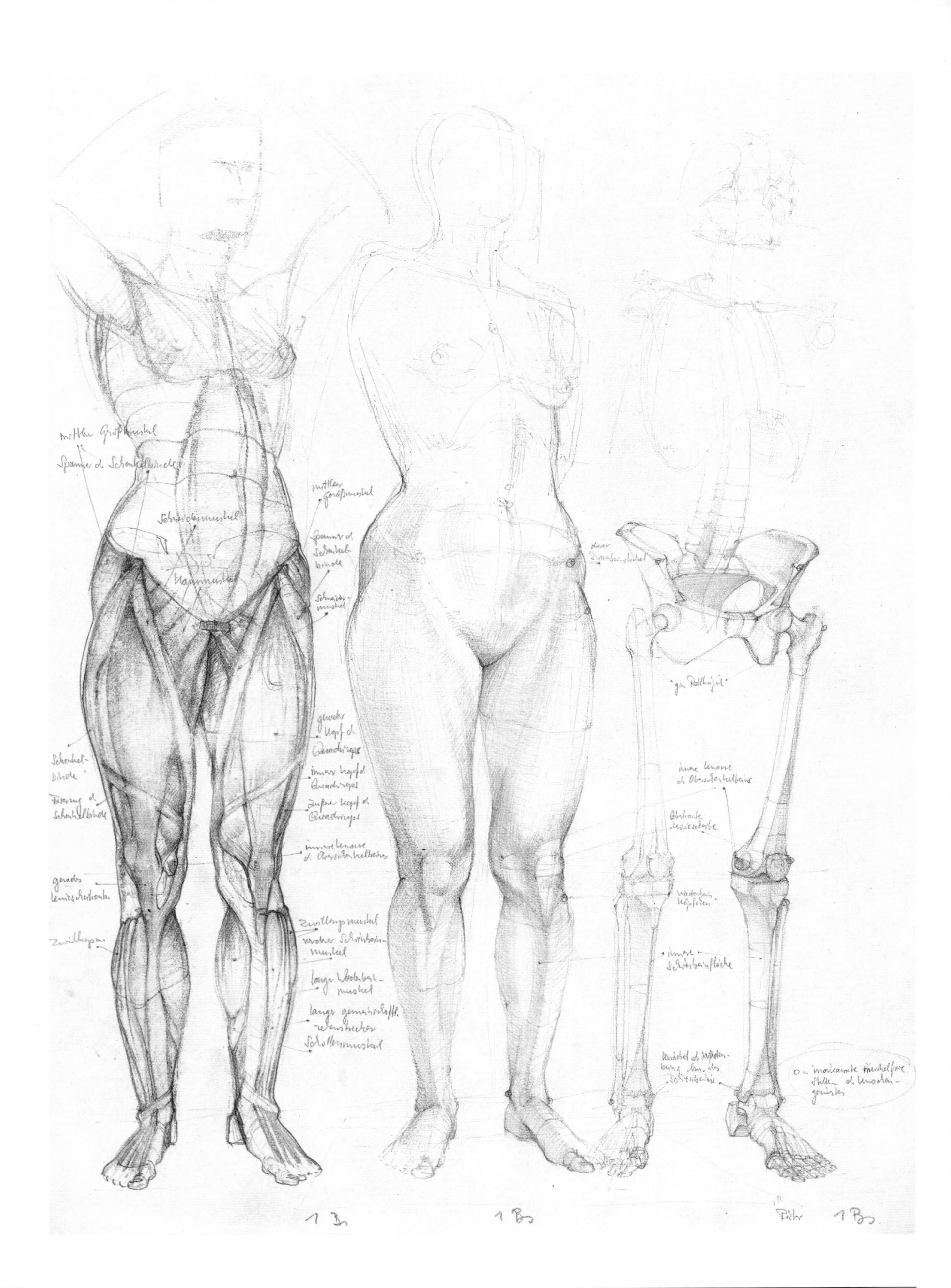

Chapter 5

Studies of the construction, function and plastic behavior of the leg

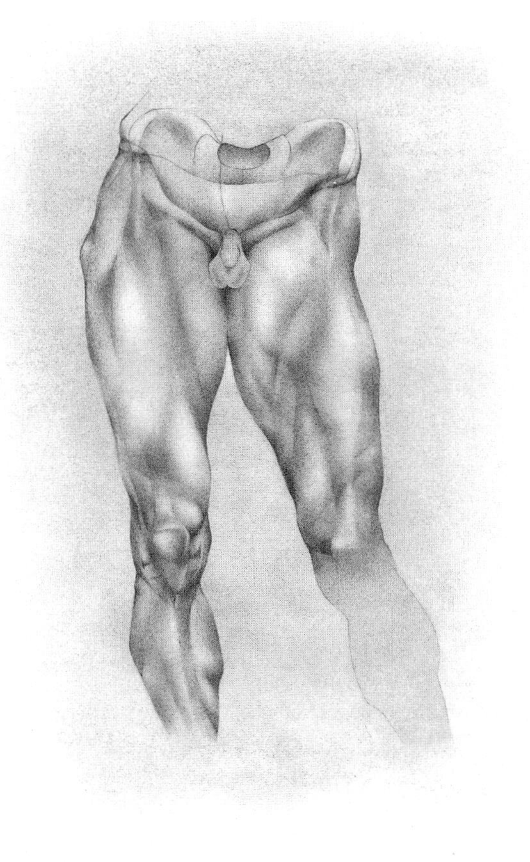

63 THREE-PART INTERMEDIATE TEST AT THE END OF THE SECOND SEMESTER
During the four-hour assignment – with no reference material allowed – the nude model appears several times so that students can prepare a complete muscle analysis, a constructional drawing of the skeleton and an architectonic life drawing of the pose.
Student of sculpture, second semester

64 RENDERING OF HIP TO KNEE
Implicit in this largely analytical way of looking at the figure is the danger of an exaggerated emphasis on certain points, but for some specialities it can be extremely useful.
Student of restoration, second semester

65 FREER TREATMENT OF THE HIP AND THIGH
Loose drawing lacking absolute precision is just as questionable as excessive attention to detail.
From a Bammes course at the Schule für Gestaltung, Zurich

66 THE SKELETON OF THE FOOT AS A CONSTRUCTIONAL BASIS FOR UNDERSTANDING ITS FORM
Thorough studies of the skeleton of the foot enable the individual elements to be comprehended as a plastic whole.
From a Bammes course at the Schule für Gestaltung, Zurich, 1988

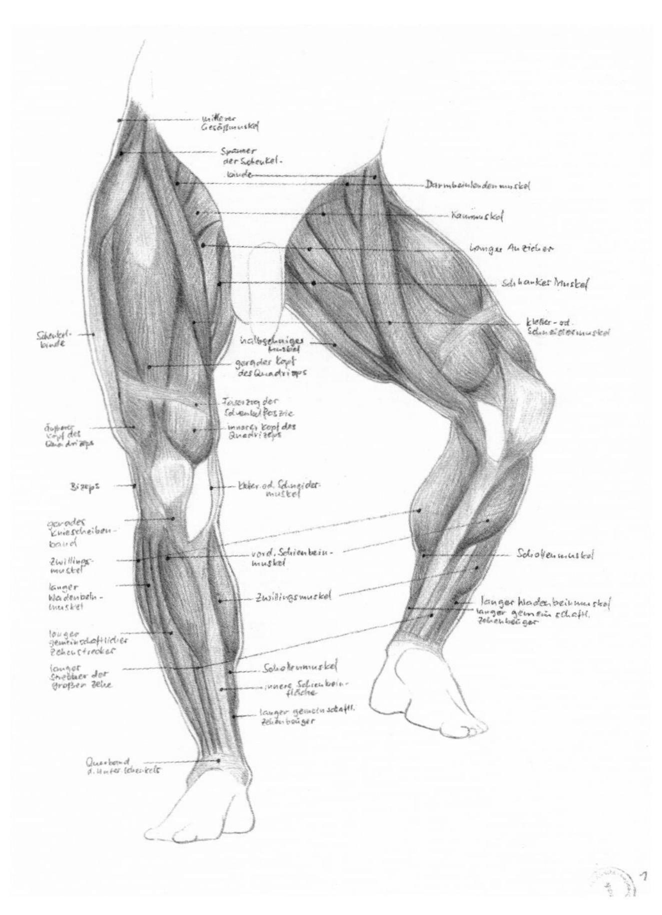

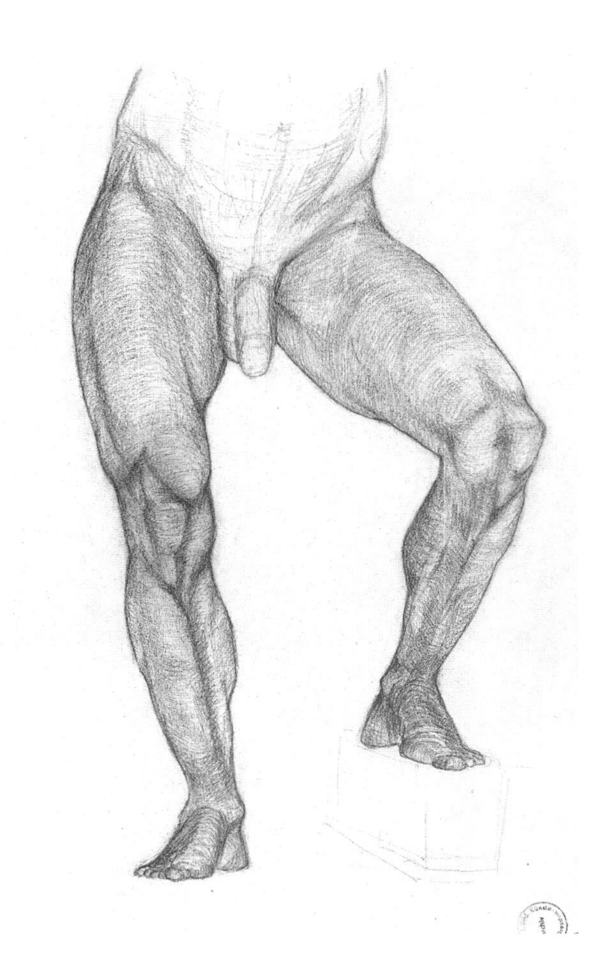

67 MUSCLE ANALYSIS AS AN EXERCISE BASED ON VISUALIZATION

As in fig. 63, it is a question of interpreting the living form concretely through analysis.

Guest student in an animated cartoon film studio, second semester

68 WORKING BACK FROM THE MUSCLE ANALYSIS TO A SYNTHESIS

Muscle analysis must never be an end in itself. Ultimately it should always be built back into a whole.

Guest student in an animated cartoon film studio, second semester

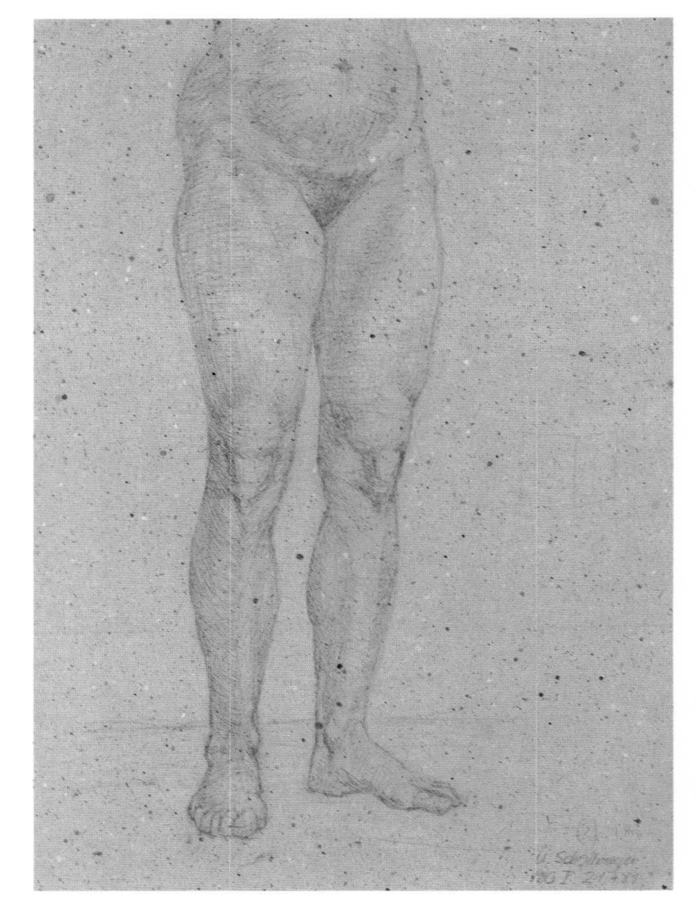

69 A WAY OF REPRESENTING THE BODY THAT IS CLOSE TO LIFE

This rendering concentrates mainly on the modeling of soft forms.

Student of painting/graphic art, second semester

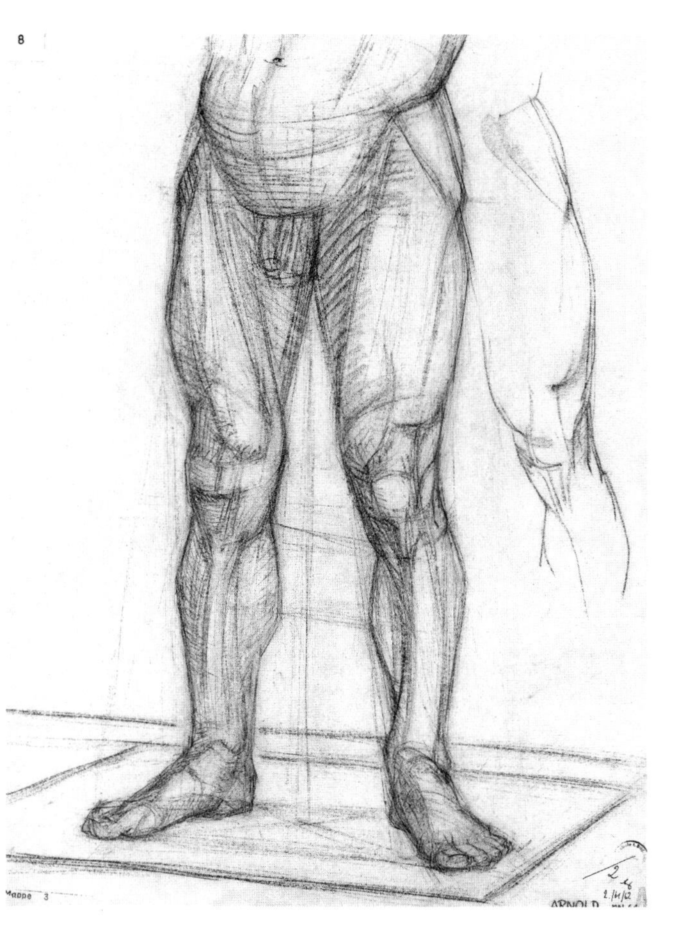

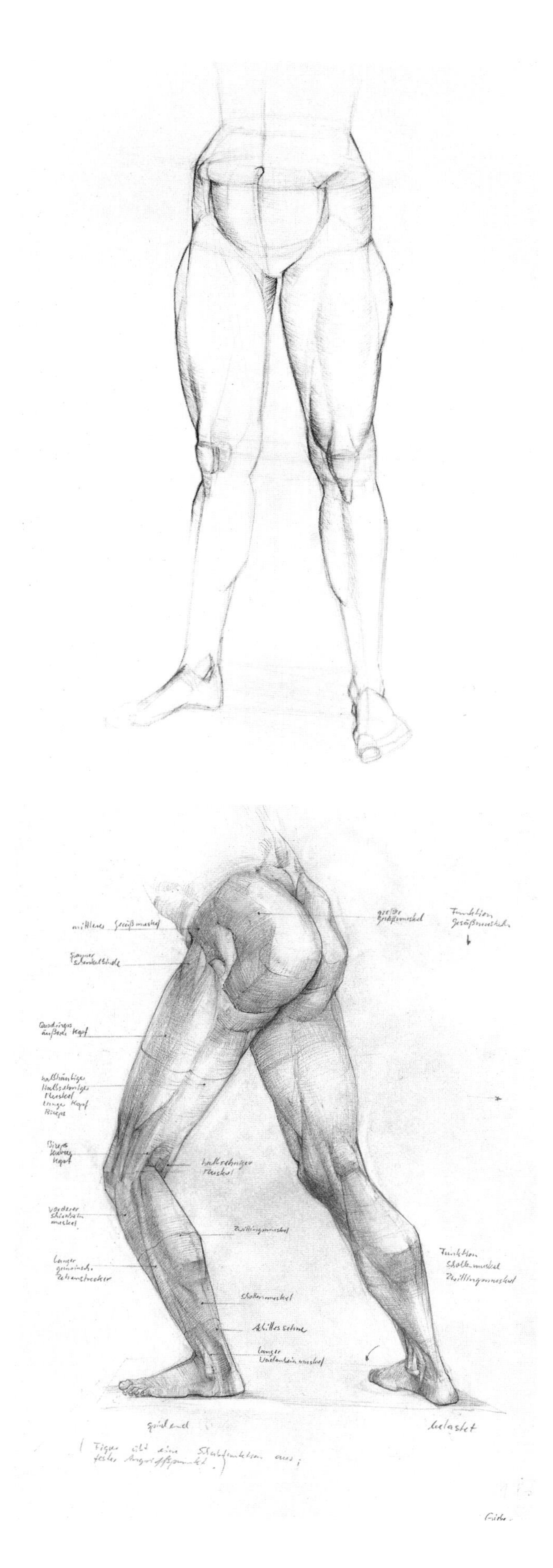

70 RIGOR OF FORM WITH CONSTRUCTIONAL ACCENTS
This rests primarily on the structural emphasis of large volumes where the forms meet.
From a Bammes course at the Schule für Gestaltung, Zurich, 1985

71 EMPHASIZING THREE-DIMENSIONAL CONNECTIONS
This is expressed in the space formed by the insides of the two thighs and particularly in the way that the continuation of the diagonal furrow on the insides of the thighs along the inner surface of the shin-bone which has no muscles has been recognized.
Student of sculpture, second semester

72 EMPHASIZING THE GROUPS OF MUSCLES BEING USED FUNCTIONALLY (TEST ASSIGNMENT)
This study examines the groups of muscles active in maintaining the pose, which also underlie the spaces formed, especially in the hollows of the knee and rear foot.
Student of painting/graphic art, second semester

73 POWERFUL REPRESENTATION OF PLASTIC BEHAVIOR OF THE BENT KNEE

We are dealing primarily with the solid, three-dimensional appearance of the two thigh-bone and shin-bone cylinders as they recede from the viewer, interrupted by the sharp-edged shape of the knee.
Student of painting/graphic art, second semester

74 CONCISELY FORMULATED STUDY OF THE LEG

This representation is confined to economical indications of the soft and hard forms.
From a Bammes course at the Schule für Gestaltung, Zurich, 1985

75 STUDY OF THE PLASTIC BEHAVIOR OF SOFT FORMS

In a sitting position the inside parts of the thighs are pressed outward, showing their 'passively' altered shapes and the two groups of opposite muscles.
From a Bammes course at the Schule für Gestaltung, Zurich, 1985

76 COMBINING PROCESSES RELATING TO STRUCTURAL FUNCTION WITH THE SENSUAL ATTRACTION OF SOFT FORMS

Experience gained in contrapposto studies and the careful handling of the behavior of muscles, fat and skin are reflected in the modeling used here.
Guest student in an animated cartoon film studio, second semester

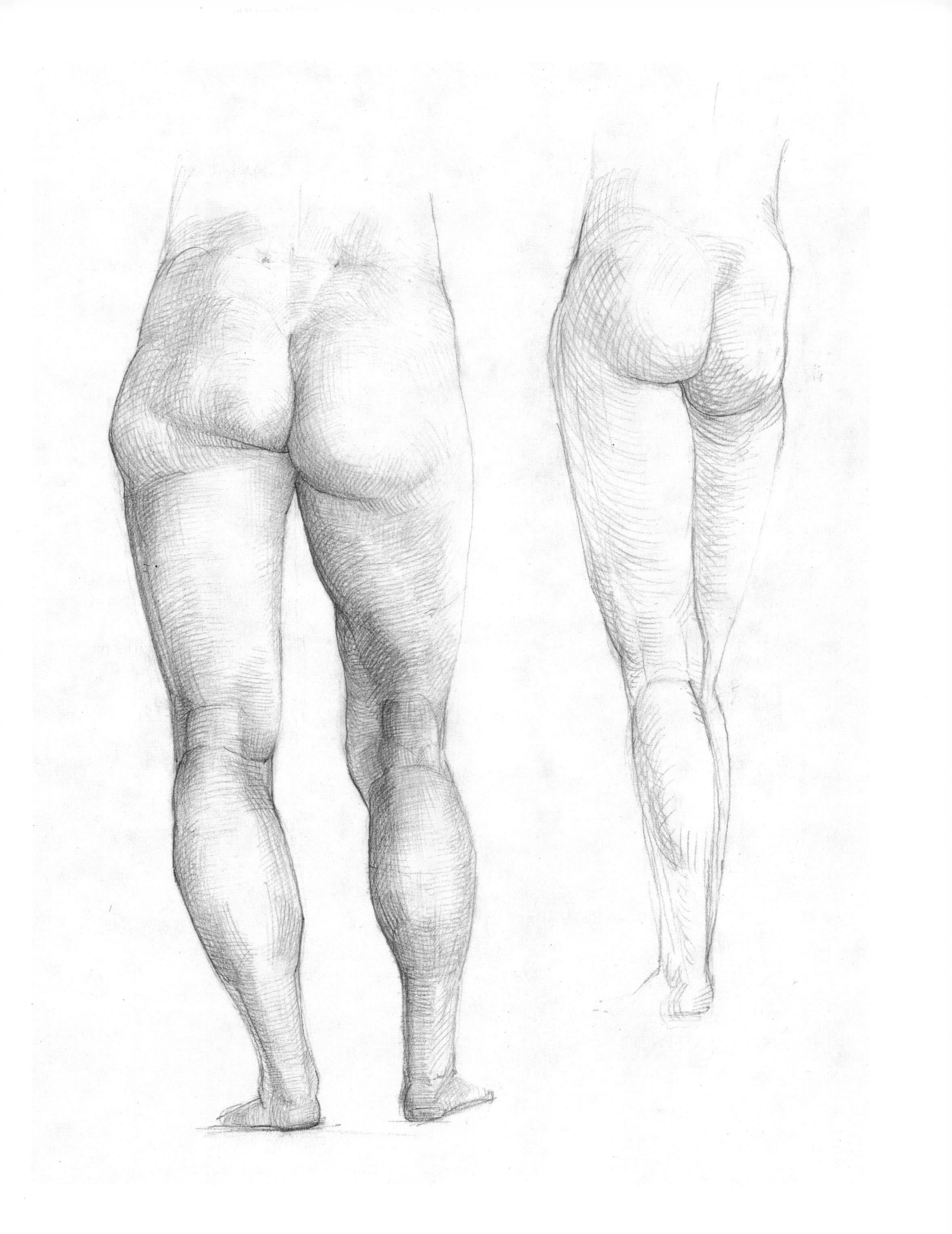

77 STUDIES OF THE PLASTIC BEHAVIOR OF THE KNEE DONE FROM THE IMAGINATION

These attempts to visualize the knee in movement from different view-points confirm that you can only draw what you have really understood, especially in the case of forms capable of such transformations as those of the knee.
Student of sculpture, second semester

78 REVERTING FROM INTENSE STUDY OF FORM TO FREE PLAY

Free expression is used at the end of a teaching session when there has been intense study involving thorough investigation of the model. The illusions using all kinds of tricks in this drawing delve into the imagination.
Student of painting/graphic art, second semester

6.

Studies of the construction, function and movement of the torso

The effects and phenomena on the torso are more pronounced than those on the limbs. On the torso thin, surface muscles have to close gaps between parts of the skeleton. Their regular pattern of behavior stems from the relationships between the plastic cores (the pelvis and the rib cage) and from the arrangement of the shoulder girdle which is the adjustable basis of the movement of the arms.

Before intelligent graphic study of the muscles of the torso can be embarked on, studies of the pelvis and rib cage (figs 79, 81) and of the structure and mechanics of the shoulder girdle (figs 79–84) must be undertaken. Unless all this preparatory work is done you cannot expect to draw the structure of the torso at rest and in motion successfully (figs 83–87). Evenually, three-dimensional studies (figs 100, 101) can be appended.

The spinal column with its double S-shape connects the different sculptural cores (figs 79, 81). The rib cage or thorax is treated as a self-contained receptacle finishing off the upper end of the torso in a dome shape that has been compressed from front to back. The following should be noted (fig. 81):

- The actual front surface is the breast bone or sternum and the costal cartilages branching off from it.
- The ends of the costal cartilages that connect with the ribs together form an outward and downward curving, slightly bowed facet as opposed to the braced form of the ribs.
- The sides of the rib cage then develop in depth (to study this, draw cross sections of the rib cage at intervals).
- The dome form opens upward in a kidney shape, widening markedly until about the third rib up, then gradually becoming more convex as it continues; for that matter all areas of the rib cage are convexly braced.
- The front section of the rib cage forms the rib arch starting from the tip of the breast bone.
- The lower opening of the rib cage is considerably wider than the upper one, but a little narrower than its widest point which is a short way above the bottom opening.

In terms of drawing, it is best to proceed as follows:

- Make a plan of the system of spatial relationships (the line followed by the central axes at the breast bone and the spinal column),
- And of the three-dimensional transverse axes at the top and bottom end of the rib cage, doing a sectional drawing of each.
- Then insert the convex forms of the dome shape along with the studies of the transverse axes into the system of reference.

The collar-bone or clavicle (figs 79, 81), which when viewed from above forms an S-shaped curve, is linked to the skeleton of the torso by a bony bridge at one point only, through its attachment to the breast-bone (sterno-clavicular joints). It therefore swings the arm, which is linked to the glenoid cavity round the sides of the rib cage and can be raised or lowered (fig. 79). At the tip of the shoulder the spine of the triangular shoulder blade is jointed with the outer end of the collar bone (figs 79, 81). The shoulder blade moves round freely over the rib cage, especially in vertical lifting or when the arms are folded on the chest, and this mobility considerably extends the arm's radius of action. The changeable relief of the shoulder girdle should be studied in detail from a nude model (figs 82, 84).

The way in which the surface of the front and side abdominal areas can extend (pure torso muscles, fig. 88) is the basis for understanding the three-dimensional effects and the alterations caused in the abdominal wall by compression (figs 93, 97), stretching (figs 83–86, 95) and twisting between the pelvis and the rib cage (figs 87, 90, 92, 94).

The soft forms of the abdominal area must be drawn in all these functional processes like transformable drapery, showing the consistent behavior of the abdominal wall and the skin. It is better to exaggerate these folding characteristics expressively (figs 85, 90, 94, 96) than to kill the effect by dull, lame, uncertain drawing.

The front and side abdominal wall extends as an intermediate form between the front and side surfaces of the rib cage and pelvis. The way the pelvis and rib cage are made is responsible for the ins and outs that occur here. Thus solid constructional drawing must always be based on imagining and remembering these two parts of the structure, so that the observer can see this and share the experience. Compared with these two main masses and optical weights all other forms – pectoral muscles, breasts, the shoulder girdle – are only subsidiary forms or layers. In the upper part of the torso we must be able to see how the rib cage dome fits up into the shoulder girdle and works with it, and the front surface and the sides must be conveyed as stable, constant forms in contrast to the shapes of the soft parts.

We must also bear in mind an important phenomenon: the shield shape of the large pectoral muscle (connecting with the arm) and the breasts lying on it. When the arm is raised, together they form a bulging cord cutting across the outer end of the collar bone (figs 83, 85, 90, 96). From the rear view the broadest back muscle (*latissimus dorsi*) extends to the inside of the upper arm, and together with the large pectoral muscle and the side of the rib cage forms the armpit. When the arm is raised vertically the deltoid muscle is inevitably compressed above the tip of the shoulder, causing a double fold to occur at that point on the shoulder bone. The gap between the adjacent deltoid and large pectoral muscles at the collar bone results in the formation of the hollow below the collar bone (fig. 89).

The three-dimensional gradations of the front of the body indicated in figs 85 and 86 must be supplemented by three-dimensional studies of the back, similar to those already shown in figs 6 and 8. The parts of the body that are farthest in the background are almost incapable of being depicted, so that they merge into space without any break (fig. 100).

79 THE THREE-DIMENSIONAL MOVEMENT OF THE SHOULDER GIRDLE

Looking down on the skeleton in an on-all-fours position gives particularly good insight into the S-shaped curve of the collar bone, and particularly into how it travels from its jointed connection with the breast bone backwards to its connection with the shoulder blade.
Student of restoration work, second semester

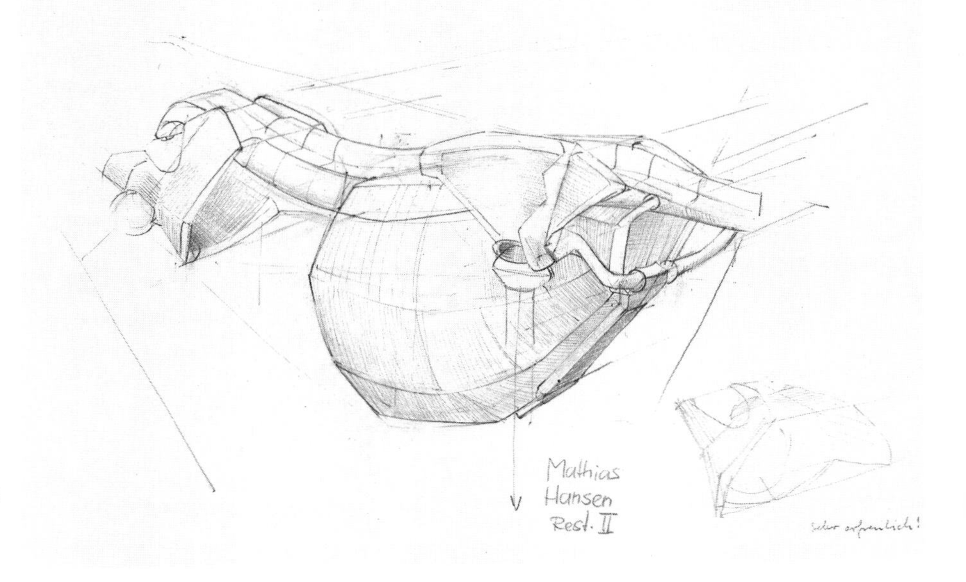

80 FUNCTIONAL STUDY OF THE FRONT PART OF THE SHOULDER GIRDLE – THE COLLAR BONE

When the arm is raised more than horizontally this causes an associated lifting of the collar bone near the tip of the shoulder; the way the pectoral muscles cut across it must be made clear.
From a Bammes course at the Schule für Gestaltung, Zurich

81 STUDIES OF THE SCULPTURAL CORES – THE PELVIS AND RIB CAGE

If you have not done a detailed study of the two containing forms, you will have difficulty in defining spatial directions when you come to do a life drawing. The shoulder girdle looks like an ornamental cladding on the dome formed by the upper rib cage.
Student of set painting, second semester

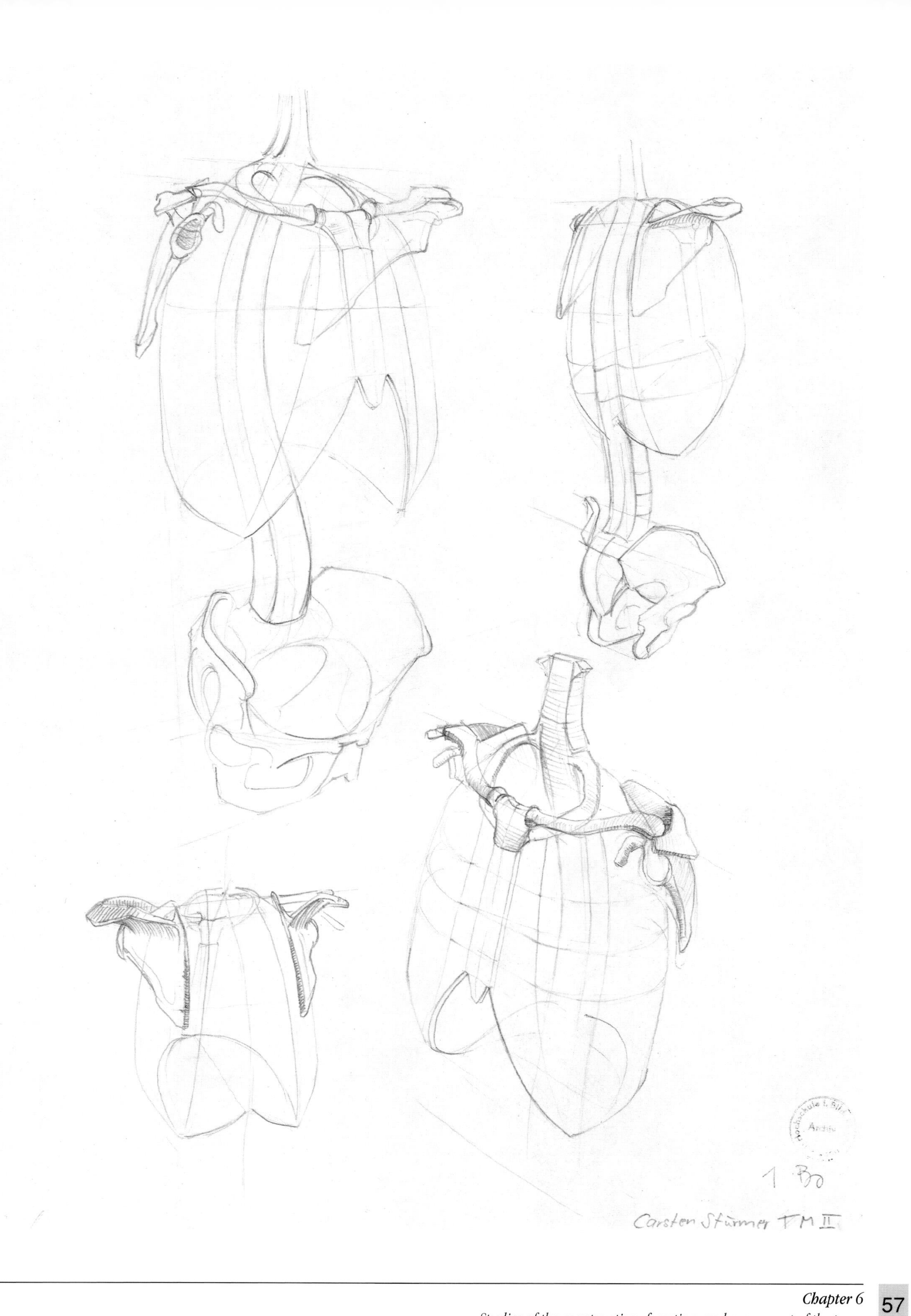
Carsten Stürmer FM II

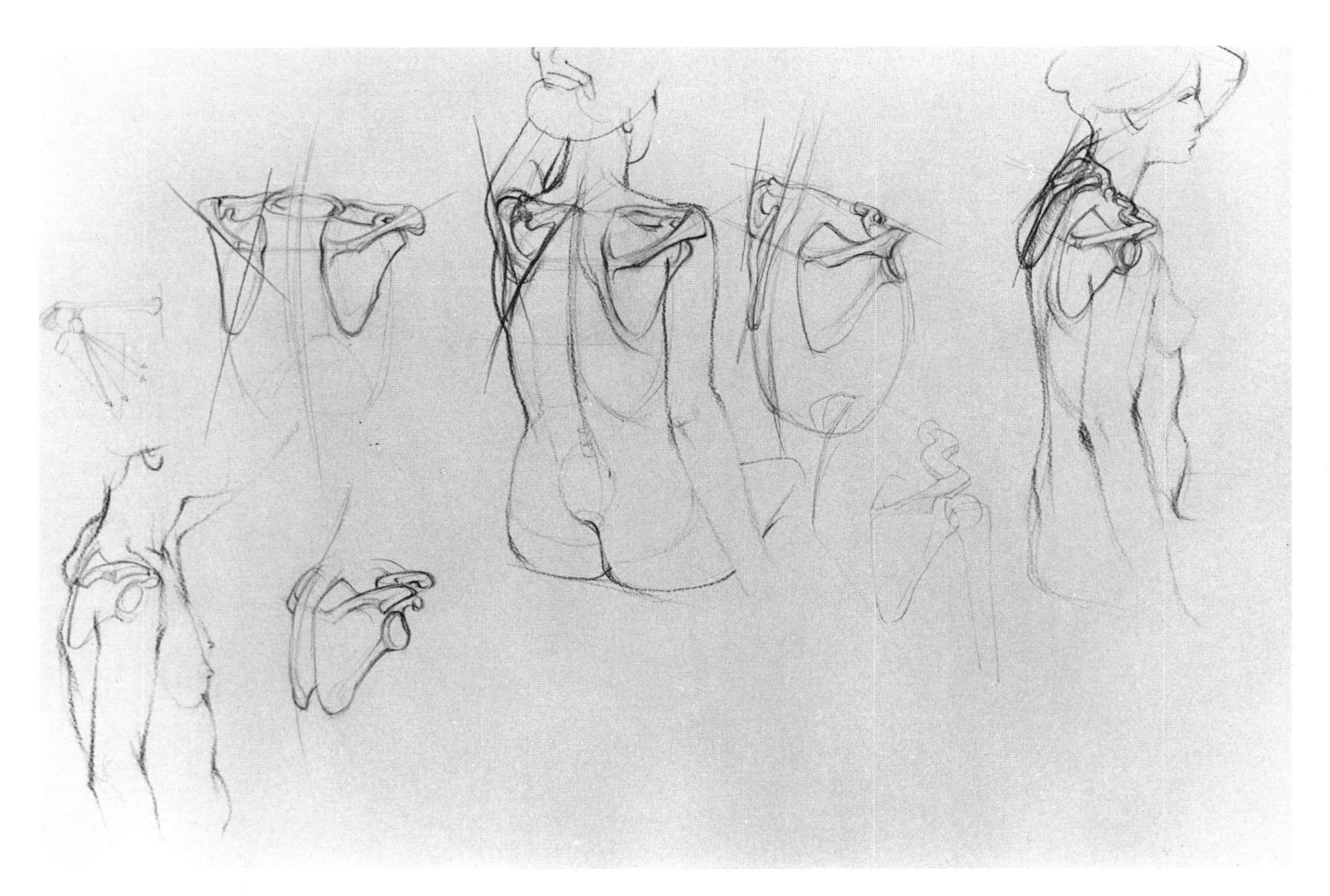

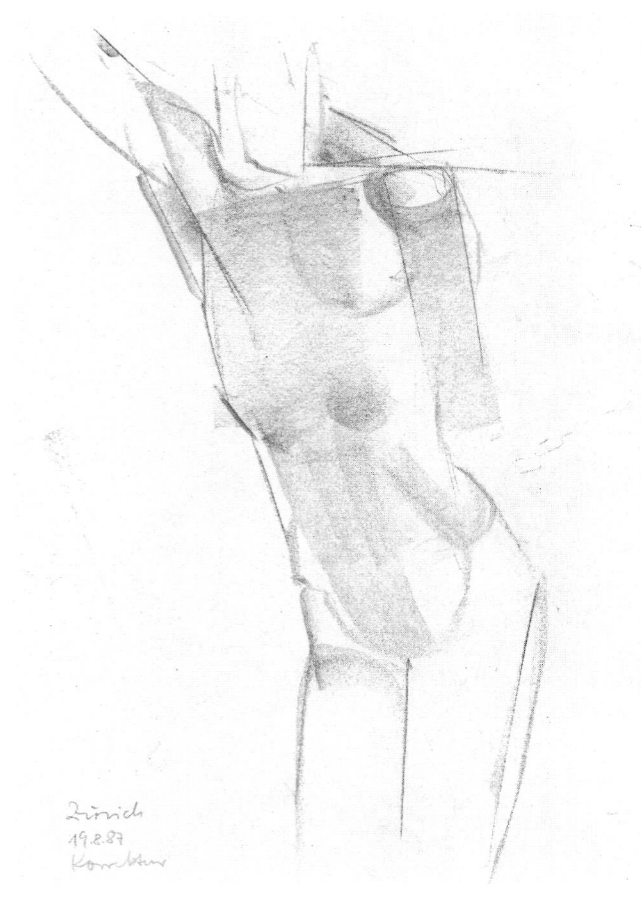

82 FUNCTIONAL STUDIES OF THE REAR PART OF THE SHOULDER GIRDLE – THE SHOULDER BLADE
Anatomically the position of the shoulder blade follows the movement of the arm.
From a Bammes course at the Schule für Gestaltung, Zurich

83 HOW BASIC FACTORS CAN BE CONVEYED
The front and side surfaces of the torso, the back-turned cylinder of the arm and the armpit area are conveyed in their complexity using broadly applied graphite stick.
Demonstration study by the author from a Bammes course

84 STUDY OF THE BEHAVIOR OF THE SHOULDER BLADE
The course of the shoulder bone, the inner edge of the shoulder blade and the position of the tip of the shoulder are used as reference points.
From a Bammes course at the Schule für Gestaltung, Zurich

85 HOW THE PELVIS AND RIB CAGE JUT OUT
In stretching, the abdominal wall is extended, causing the cores – the pelvis and rib cage – to jut out.
Demonstration study by the author from a Bammes course

86 THE THREE-DIMENSIONAL TORSO
Basic three-dimensionality can be conveyed by dipping your finger into powdered graphite and producing smudges with it.
Demonstration study by the author from a Bammes course at the Schule für Gestaltung, Zurich

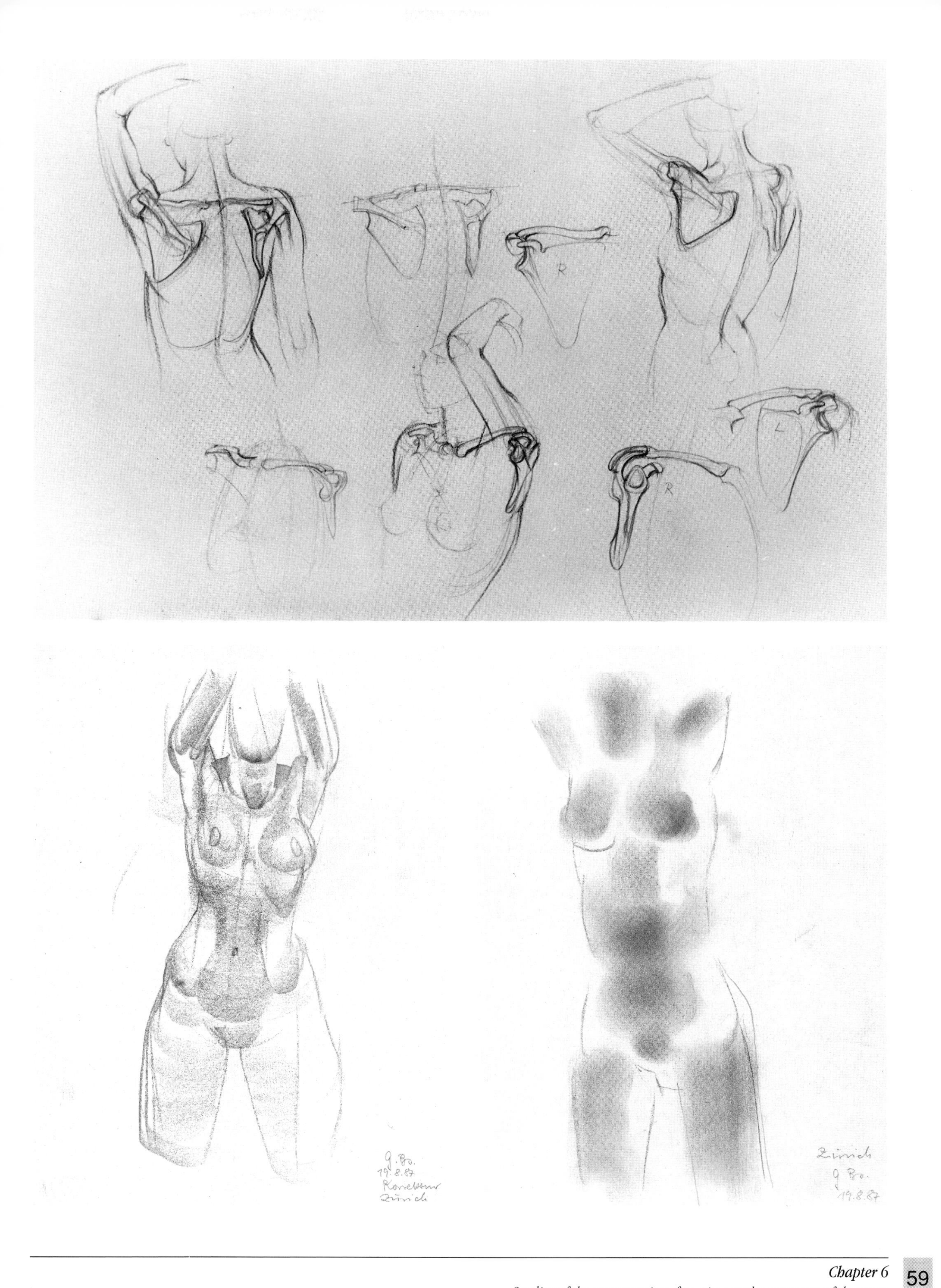
R
L
R

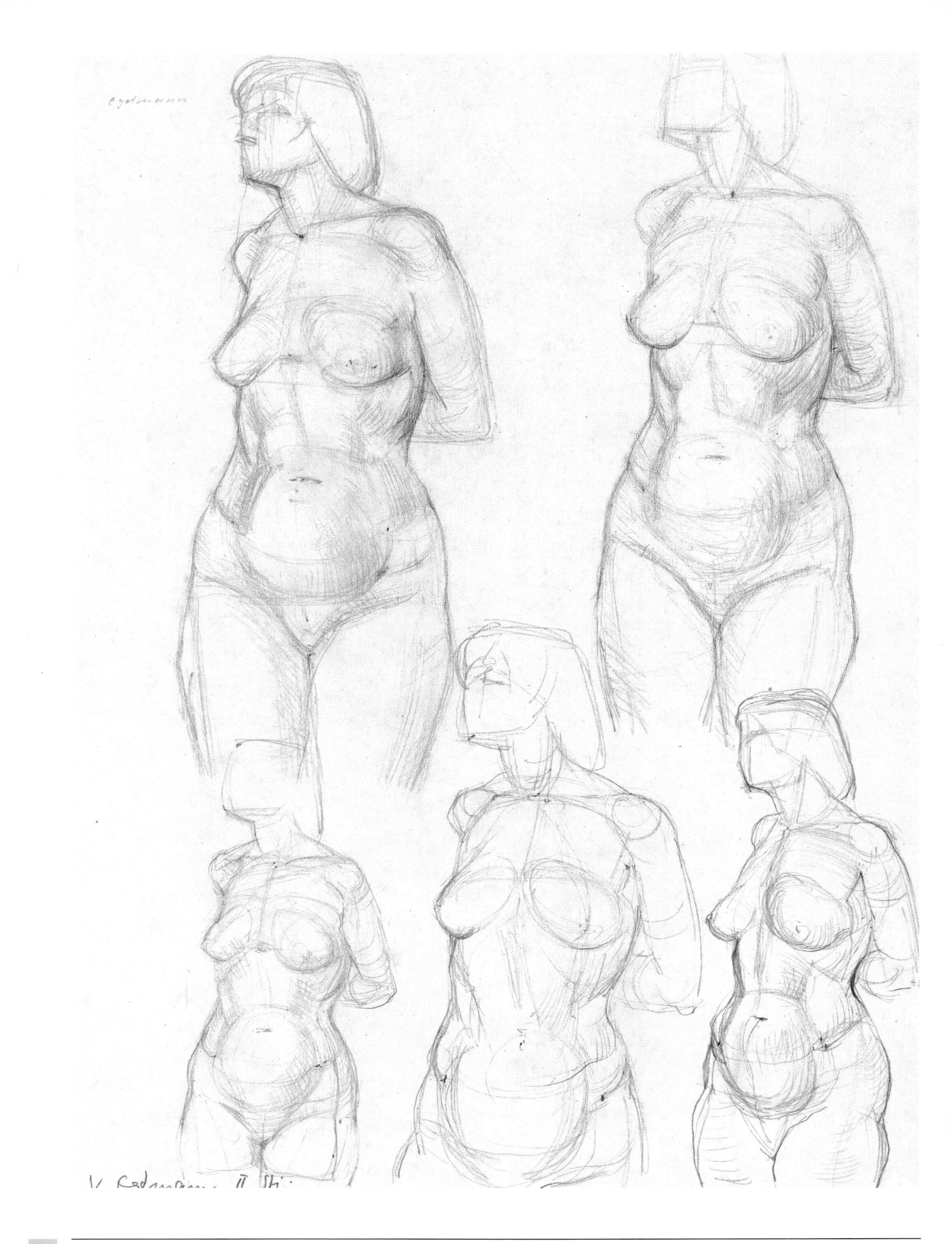

Chapter 6

Studies of the construction, function and movement of the torso

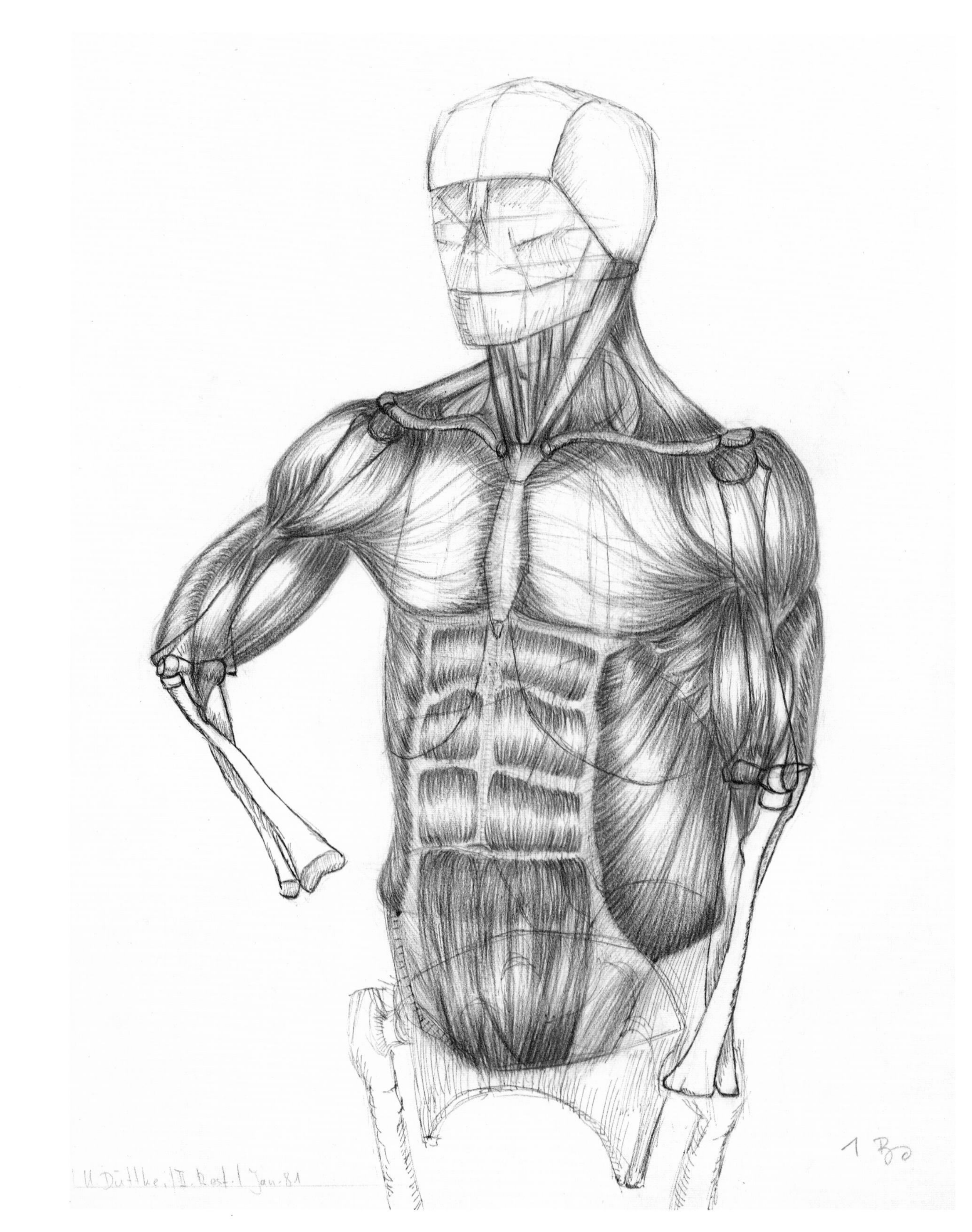

87 WHAT HAPPENS TO THE TORSO IN TWISTING MOVEMENTS

The altered position of the rib cage above the pelvis causes spiral lines on the abdominal wall.

Student of sculpture, second semester

88 ATTEMPT AT RECONSTRUCTING THE MUSCLES OF THE TORSO

A student's knowledge of the positioning of the muscles of the torso is demonstrated in an analysis based on a nude model.

Student of restoration, second semester

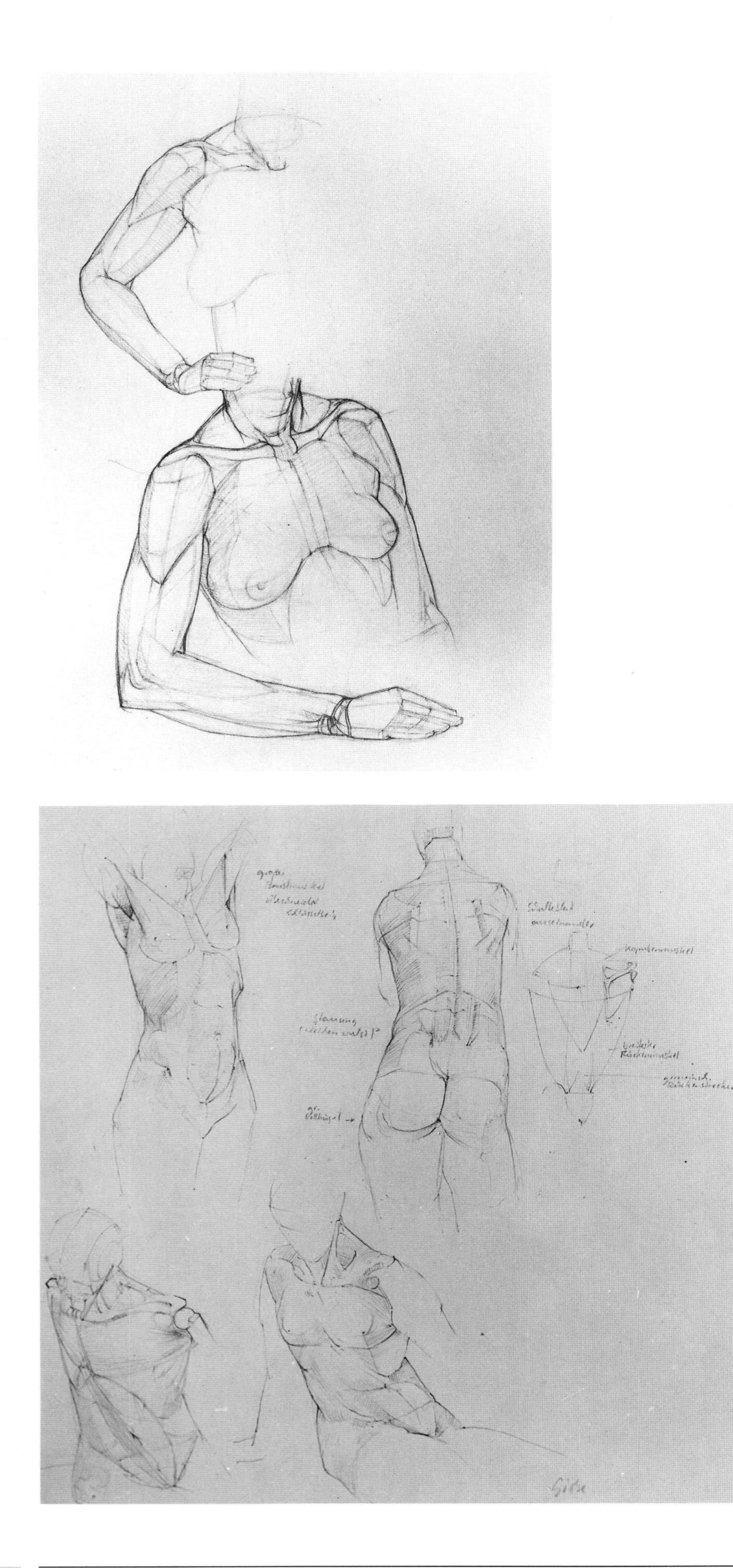

89 RELATION OF THE ARM TO THE PARTS OF THE UPPER BODY

In the upper body bony structures alternate with soft forms. Carefully observed intersections give information about what is happening three-dimensionally, particularly about where the arm starts.

From a Bammes course at the Schule für Gestaltung, Zurich

90 FUNCTIONAL STUDIES OF THE BEHAVIOR OF THE ABDOMINAL WALL

If you have a good grounding in the shape of the pelvis and rib cage, you will find it easy to define the consistent behavior of the soft forms resulting from changes in the relationship between the pelvis and rib cage.

Student of painting/graphic art, second semester

91 THE MODELED FORMS WITH SOME ATMOSPHERIC EFFECTS

This drawing breaks away from strict formal construction, using atmosphere and surface effects. The free, light drawing technique produces a lively and impressive graphic presentation.

From a Bammes course at the Schule für Gestaltung, Zurich

92 STUDY OF THE PLASTIC IMPACT OF FUNCTION ON THE MODELING OF THE ABDOMINAL WALL AND THIGH

This drawing concentrates on the abdomen suspended between the pelvis which is in a reclined position and the supported rib cage. The thighs are in a relaxed position one above the other, and between them and the abdomen there is a deep groove.

Student of set painting, third semester

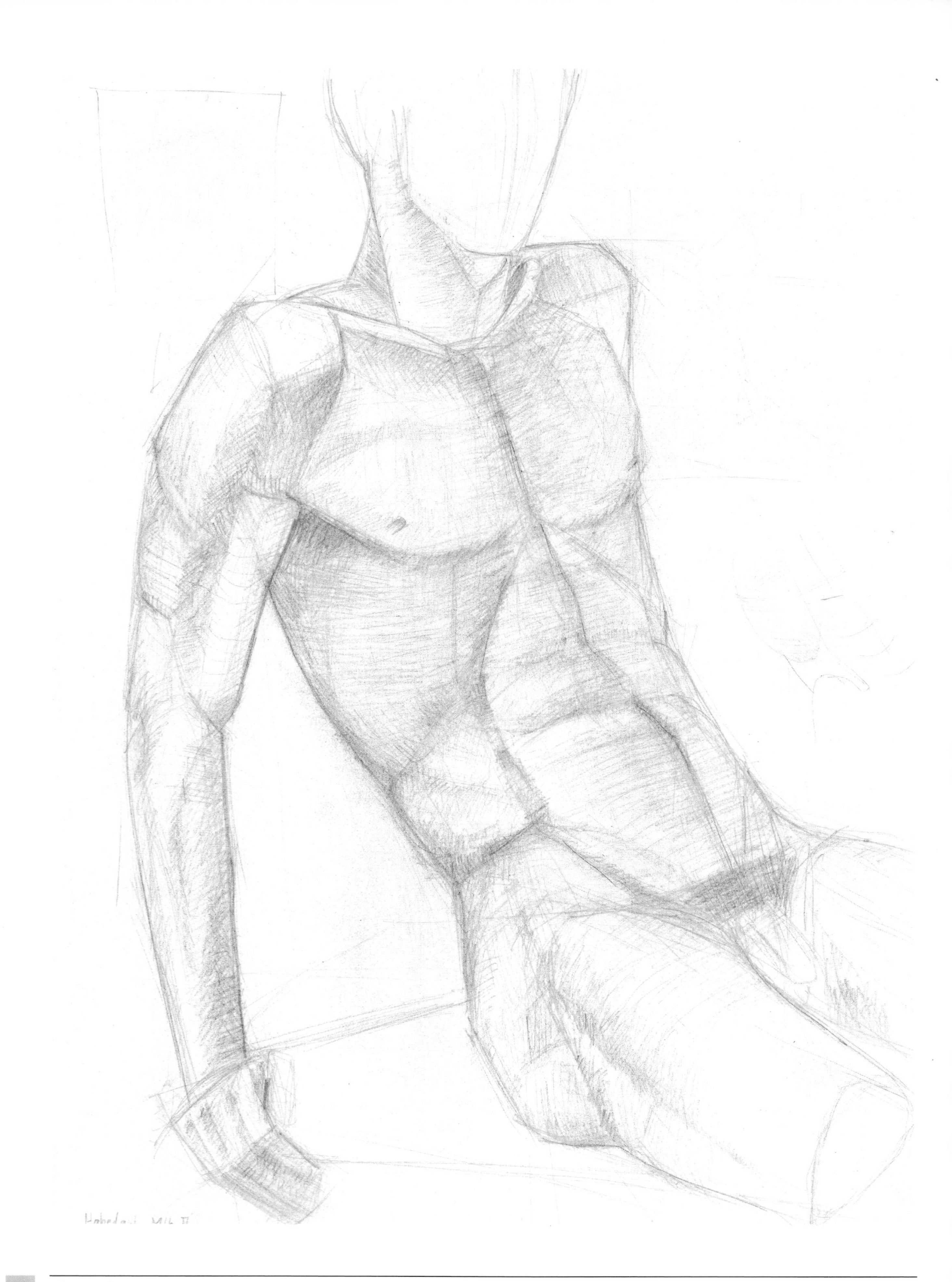

93 TWO FOCAL POINTS IN THE INVESTIGATION OF FORM

The supported backward-leaning upper body concentrates attention on the behavior of the shoulder girdle which sinks down between the two columns formed by the arms, and on the crosswise folds on the stomach wall caused by compression.
Student of painting/graphic art, third semester

94 STUDY OF SIMULTANEOUS FUNCTIONAL EFFECTS

The altered levels of the pelvis and rib cage in a twisted position have been worked out, as have the diagonal pull on the abdominal wall associated with them and its simultaneous sideways compression and stretching as a result of being supported.
From a Bammes course at the Schule für Gestaltung, Zurich

95 SHAPES CONSISTENTLY FORMED IN A SEATED, STRETCHING POSE

Note how the pelvis remains vertical in a seated position so that the forward curvature of the spine in the lumbar region is flattened, and how the rib cage and breasts rise and the stomach wall is stretched when the arms are raised.
From a Bammes course at the Schule für Gestaltung, Zurich

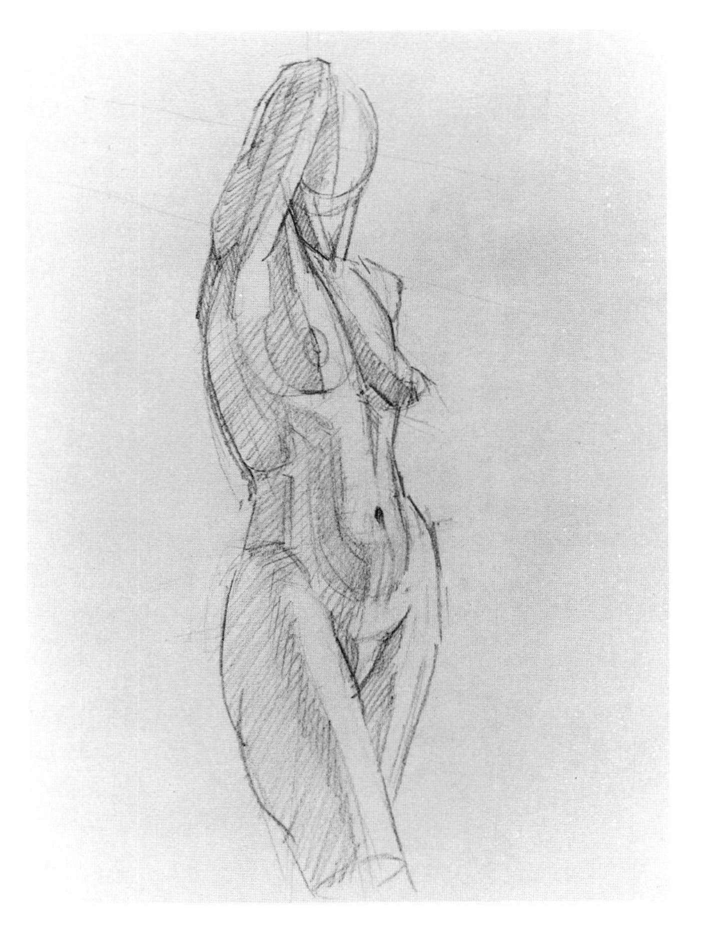

96 HIGHLIGHTING THE BODY'S CONSTRUCTION WITH OBLIQUE LIGHTING

The straight abdominal muscle stands out as a volume through its gradation with the slanting stomach muscle beside it. The side of the rib cage and the stomach muscles combine to form the profile of the torso.
From a Bammes course at the Schule für Gestaltung, Zurich

98 FREE GRAPHIC TREATMENT OF THE UPPER BODY

Fine hatching and contour lines create more of an expressive approach than an organized study of forms and processes.
From a Bammes course at the Schule für Gestaltung, Zurich

99 THE ARCHITECTURE OF THE TORSO USING PURELY LINEAR DRAWING

The fundamental layout of the three-dimensional reference system can be clearly recognized, especially the three-dimensional, rhythmic course of the line running down the center of the body. The positions of the basic volumes of the pelvis and upper body are oriented to it.
From a Bammes course at the Schule für Gestaltung, Zurich

100 STUDIES OF THE TORSO CONCENTRATING ON THREE-DIMENSIONALITY

The solid volume of the body is suggested just by conveying the three-dimensional gradations from the front of the body to the parts farthest from the viewer. The dialog between three dimensional and solid appearance must always be present in drawing.
Student of painting/graphic art, third semester

97 THE RELATIONSHIP BETWEEN STRETCHING AND COMPRESSION

The way the rib cage is drawn down toward the pelvis in a forward-bending sitting position compresses the abdominal wall, and the back is under tension. The arm is brought forward, pulling the widest back muscle (*latissimus dorsi*) forward with it.
Student of painting/graphic art, third semester

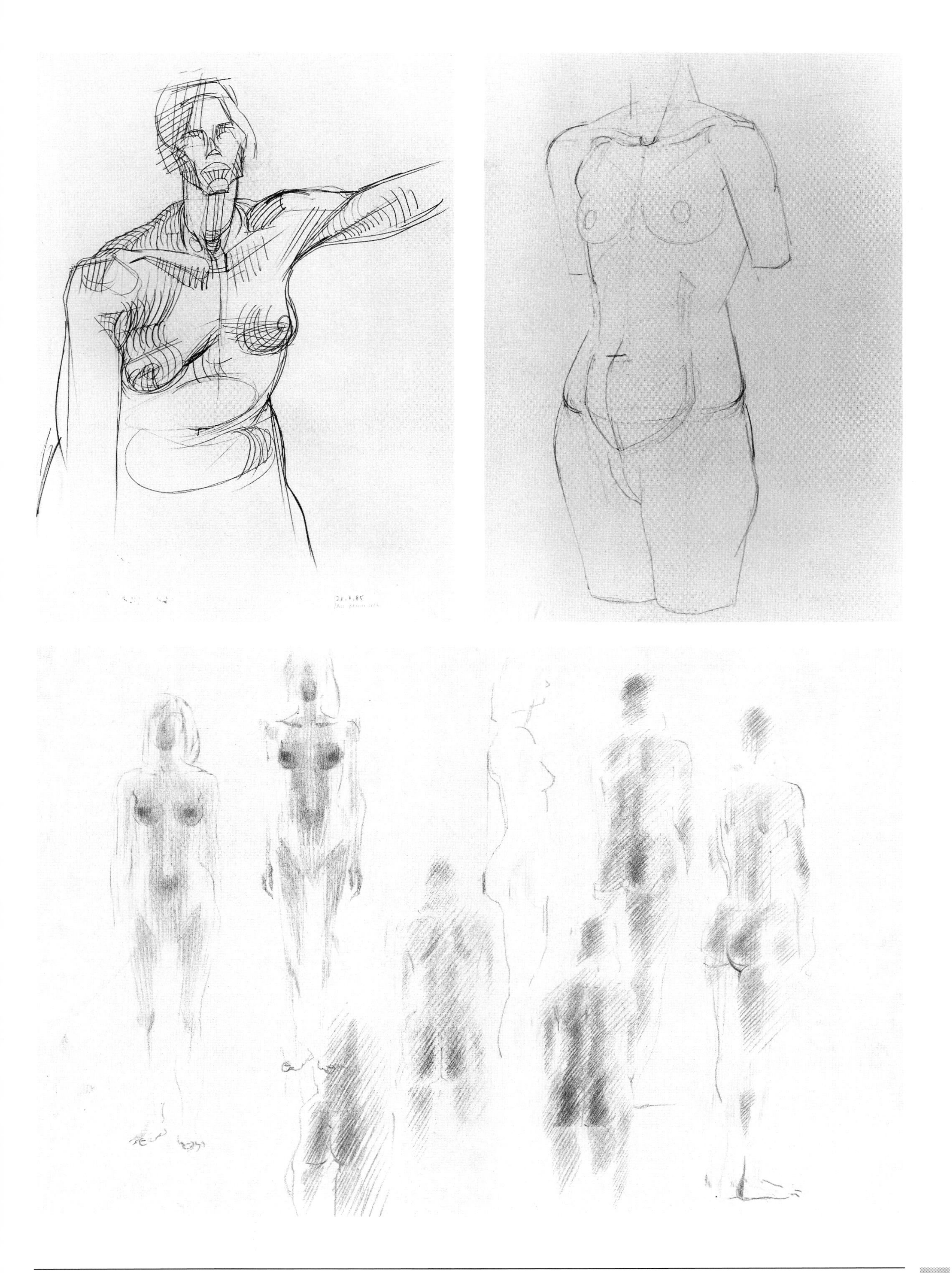

101 SOLID AND THREE-DIMENSIONAL INVESTIGATION IN HARMONY

All endeavors to establish anatomical correctness of form must ultimately be combined and enhanced in a synthesis aiming at completeness.

Student of painting/graphic art, third semester

7.

Studies of the skeleton based on visualization

So far, with the exception of the skeleton of the arm and the hand, we have always examined the wider context of the body when studying individual sections of the body from life. However, such studies are indispensable stages, in working from life, toward understanding the structure of the live human form. By such studies we do not just mean the inside of the body which is remote to the eye and which is represented in the vessel-like character of plastic cores such as the skull, pelvis and rib cage. These container shapes also have the special characteristic of being able to be seen in conjunction with the external modeling of the body

As well as such fundamentals as spatial structure, the depths of the body also conceal skeletal layout such as the shapes of joints; only through studying them and the muscles is it possible to understand mechanical processes. In our emphasis on constructional forms we have aimed to promote understanding of the essentials of organic forms.

In setting out to equip students of drawing with some knowledge of anatomy, it is intended not only to give them basic knowledge and graphic skills relating to organic living objects, but also to enable them to construct figures based on visualization; producing drawings of the skeleton based on conceptualization is particularly important in achieving this.

The shape of the body understood architectonically undoubtedly provides important information about changing structural appearance in living forms, and the ability of the inner eye to visualize the form and behavior of the body's framework – the key to architectonic structure – is just as essential.

My experience as a teacher has confirmed again and again that studies of the individual parts of the skeleton from life constantly need practicing by means of work based on visual imagination. This makes greater demands on students, but also increases their freedom of movement, for they are forced to assemble the mental images they have formed of the construction of the individual forms into a larger whole and let the interplay of the parts within the whole emerge (figs 102–109).

This is one of the important ways of checking what has been well learned, providing evidence as to whether abilities and skills have been successfully absorbed or not yet acquired.

Tests can be set that cover a range of degrees of difficulty, whether in the form of a 'supplementary' analysis of the skeleton from a nude model, or the free invention of movements. Both forms of investigation are presented in this chapter, analyses after a model in figs 102–106, and free visualizations of skeleton movements in figs 107–109.

All these examples demonstrate the skeletal system in action, either through skeletal analyses from a model or in free visualizations of movement, and show that the jointing and foreshortening processes involved must have been understood. In analyses after a model the most important principles are the following (figs. 103–106):

- Use of the interplay of function and construction in action, particularly in regard to the shapes of joints, showing it has been understood. Graphic portrayal of the simplified skeletal forms must carry 'conviction', i.e. the simplification of the forms must reflect the capability of the joints, and what functional roles they are in a position to carry out.
- The convincing expression of movement inherent in the model's pose must be mirrored in drawing the constructional skeleton forms.
- The three-dimensional considerations involved, i.e. which parts of the body are nearest to and farthest from view, must again be clearly shown in the skeleton.

The same principles apply to free visualization of movement. This is a more difficult exercise because it entirely dispenses with the physical presence of a model (figs 107–109):

- You should not invent just any type of 'movement'. An invented movement not based on an accurately visualized activity turns the figure into a jumping jack.
- So you should visualize a quite specific activity: standing in a specified context, e.g. sitting (fig. 107) or lying. The range of movements connected with sport and work is inexhaustible (figs 107, 108).
- To achieve a powerful expression of motion, you must involve yourself fully in the activity, empathize with it, imitate it; ideally the person drawing should try to imitate or approximate the activity with his or her own body. Since it is not just a question of motion governed by the mechanics of the body, but also of psychological emotion, you have to attune yourself to it spiritually.

The degree to which you decide to simplify or supply detail will vary from person to person. Expressing forms as individual elements is part of the process of assimilation. The organization of volumes and their relationships with one another can be seen as criteria by which to judge whether the drawing makes sense if the parts are reduced to elements. Where differentiation has been developed, it is important to adhere strictly to the hierarchy of forms; the intricacy of the interlocking skeletal system requires the especially careful consideration of the volumes of the smaller joints, their connections with the parts depending on them and with the all-embracing whole.

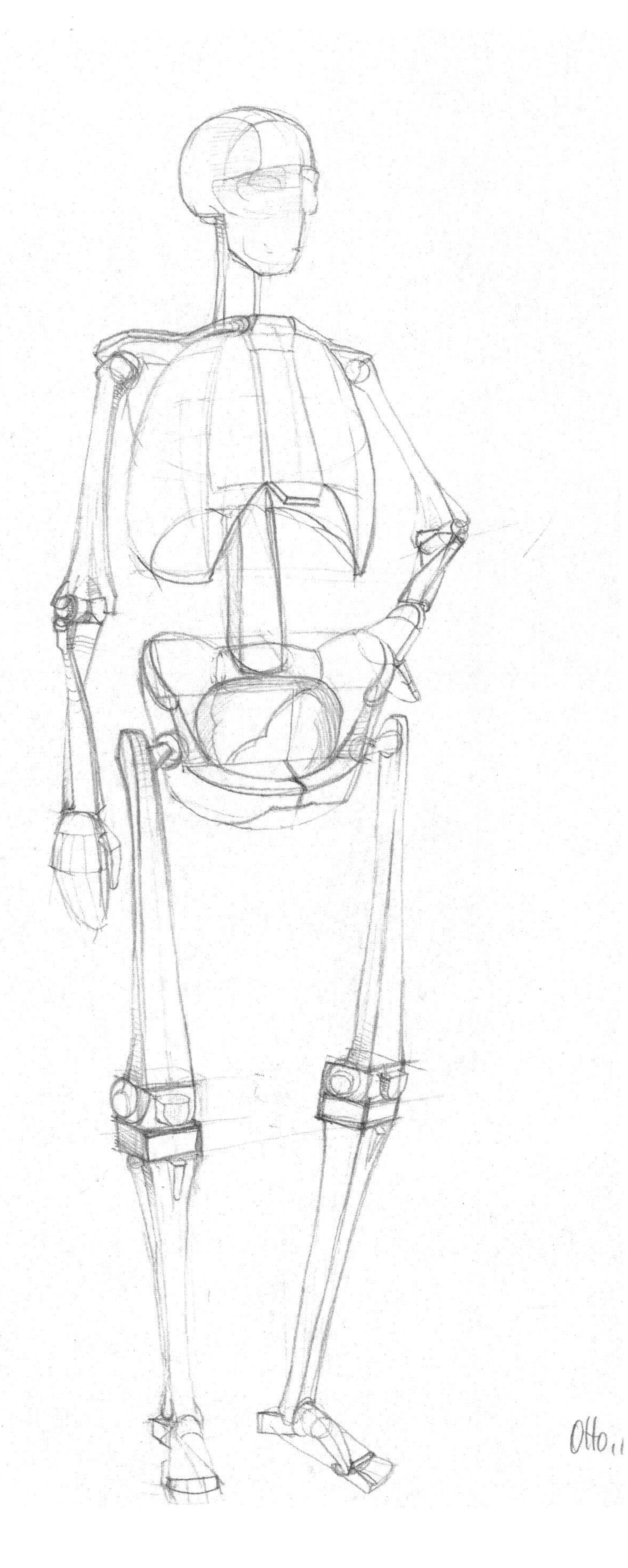

102 THE CONSTRUCTIONAL SKELETON FORMS IN A SIMPLE STANDING POSITION

As in every other study of the body, a clearly defined viewing angle is important, in terms of which the model's pose can be analyzed down to its skeletal basics. This will reveal where and how gaps in the student's ability to visualize constructionally occur.

Student of painting/graphic art, third semester

103 THE CONSTRUCTIONAL SKELETON FORMS USED TO ANALYZE A SEATED POSITION

The endeavor to clarify the relationship and behavior of the plastic cores (position of the pelvis and rib cage) and the exaggeration of the limbs jutting forward toward the viewer are clearly discernible.

Student of stage design, third semester

Mosler, Bühne II

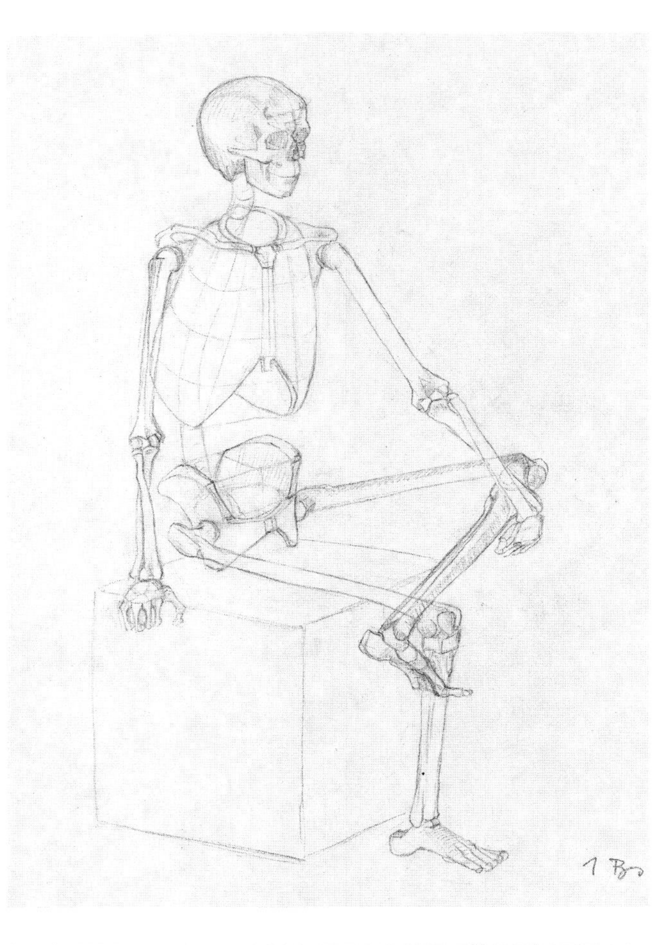

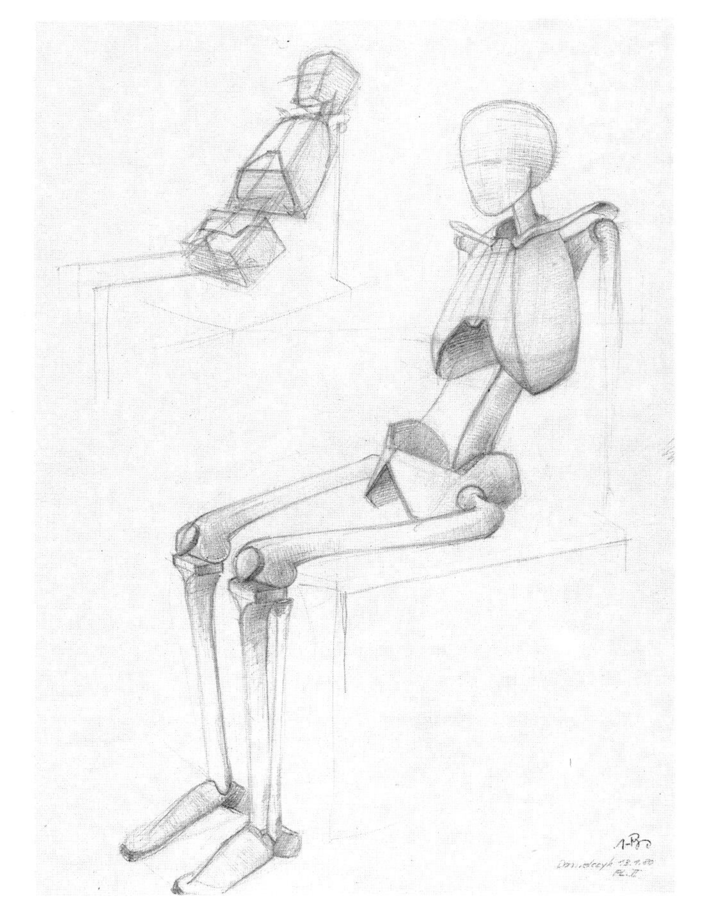

104 SKELETON ANALYSIS SHOWING CONSIDERABLE DETAIL AFTER A MODEL'S POSE

Development of the skeleton forms and their functional processes, especially in the area of the crossed legs, is central to this work.

Student of stage design, third semester

105 SKELETON ANALYSIS IN WHICH FORMS ARE CONSPICUOUSLY SHOWN AS INDIVIDUAL ELEMENTS

The shape and interrelationship of the sculptural cores (see also the subsidiary drawing) in a backward-leaning seated position are centered on the area of the shoulder girdle where the skeleton of the torso is suspended between the supporting arms.

Student of sculpture, third semester

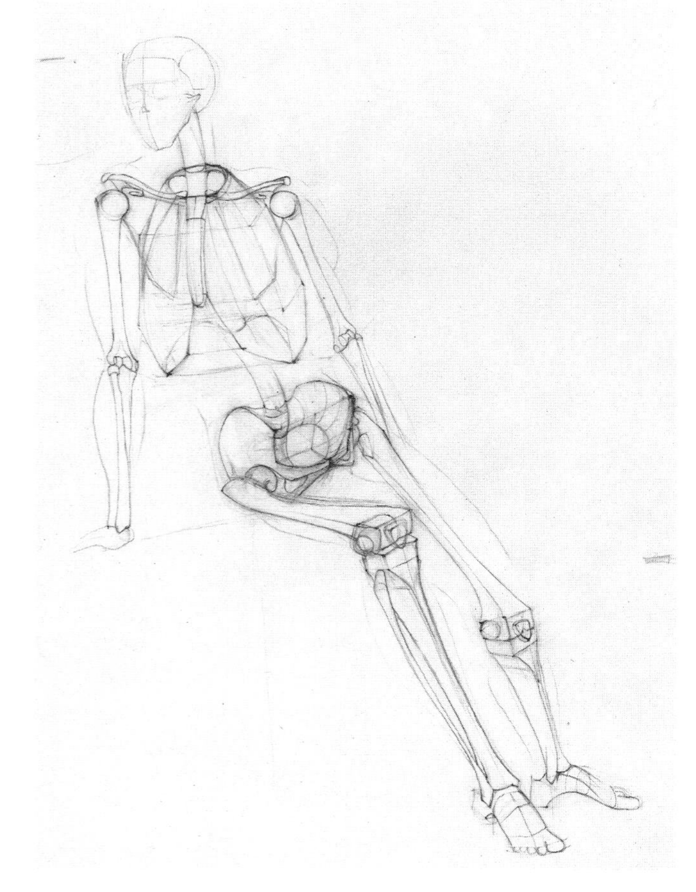

106 SKELETON ANALYSIS AFTER A MODEL IN A SEATED POSE WITH MULTIPLE FUNCTIONAL ASPECTS

The problem lies in resolving what happens when the body is supported on one arm with the pelvis bearing the load on one side, the pelvis and rib cage going in opposite directions, and the bent leg resting on the floor.

Student of stage design, third semester

107 FREELY CONCEIVED MOVEMENTS FOCUSING ON EXPRESSION OF MOVEMENT

The skeleton is shown in very different activities; the ability to convey expressive conviction comes from the very precisely visualized functional processes.

Student of painting/graphic art, third semester

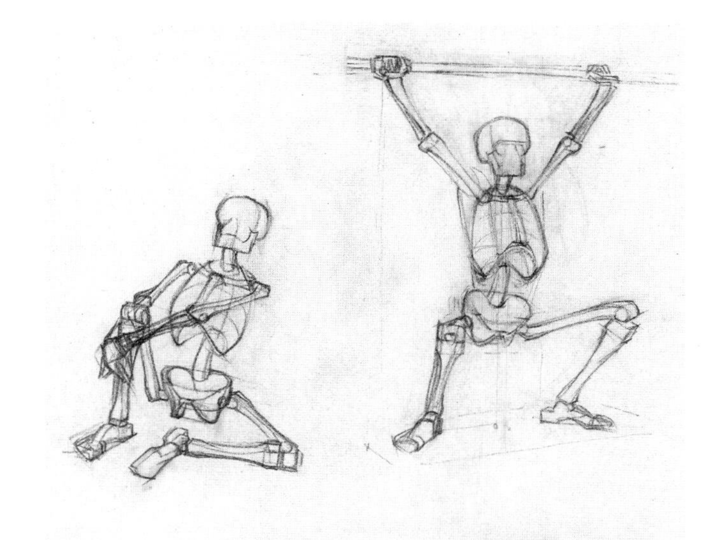

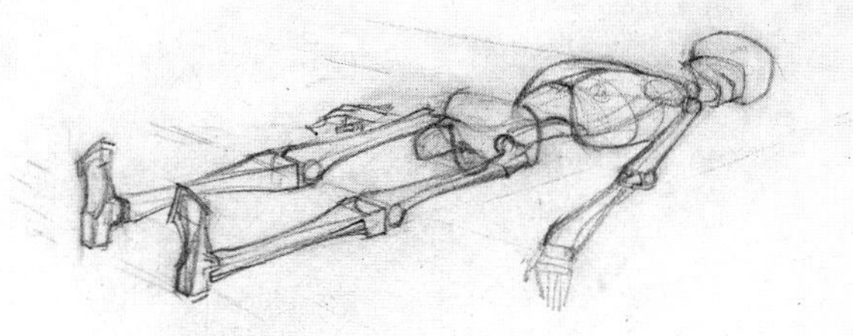

108 FREELY CONCEIVED MOVEMENTS WITH POWERFUL FORESHORTENING CHOSEN BY THE STUDENT

The successful drawing of difficult skeletal foreshortenings largely depends on the ability to visualize, and the constructional shapes of the skeletal framework and joints follow on from it.

Student of restoration, third semester

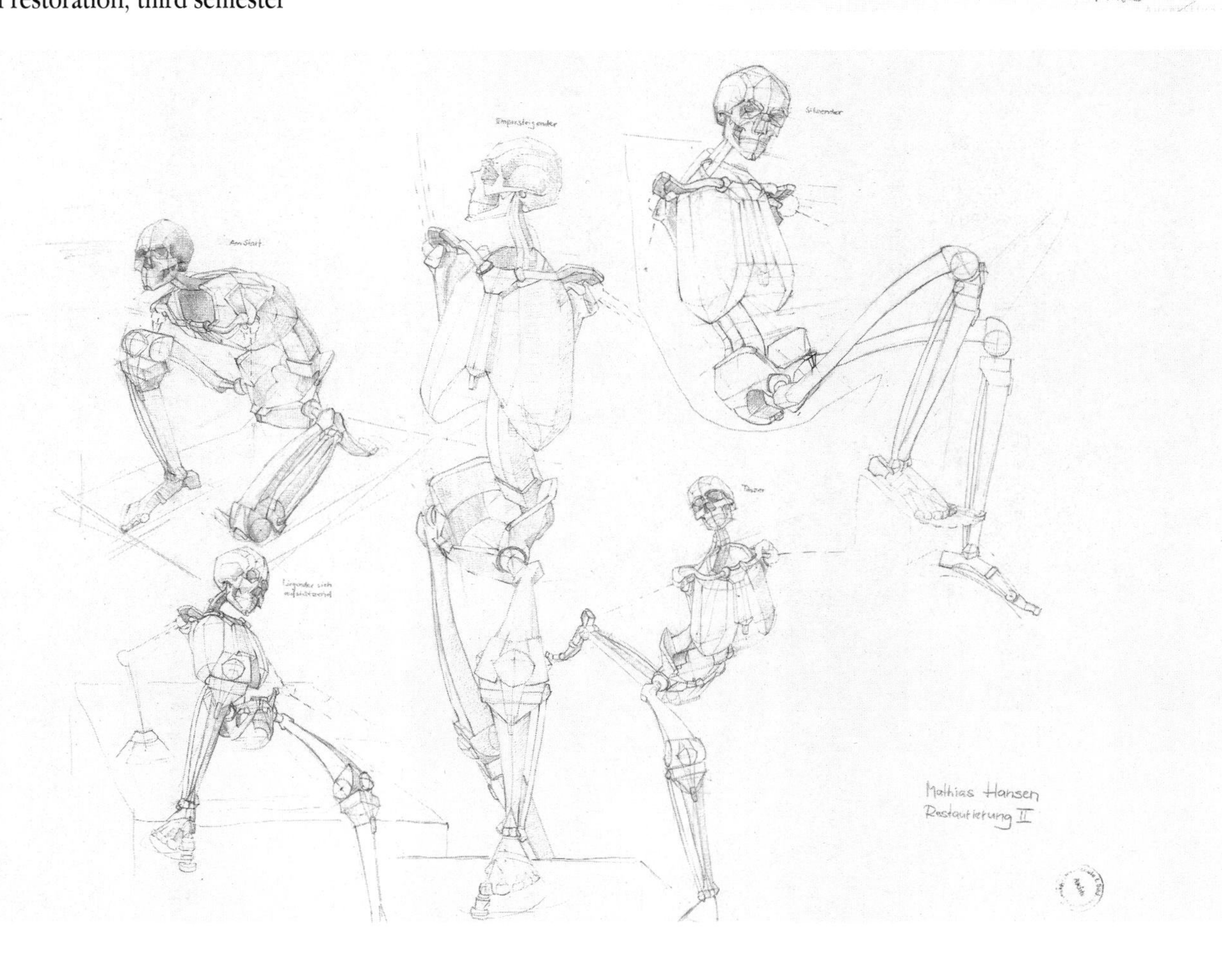

109 FREELY CONCEIVED MOVEMENTS EXECUTED AS A PROVISIONAL FINAL TEST

All the drawings of skeletons in figs 102–109 are excellent examples of work achieved under the stimulus of visualization. In the studies below the student was allowed to omit the skeleton of the arm which had not yet been covered, so that the testing of ability to draw the skeleton from the visual imagination would not be delayed.

Student of restoration, third semester

8.

Studies of the construction, functioning and plastic behavior of the hand and arm

Perceptive and accurate studies are of vital importance when it comes to the skeleton of the foot, arm and hand. The delicacy of the construction of the arm and hand and the extreme differentiation and functional universality associated with it are still touchstones of an artist's ability to convey functional and gesticulatory expression in graphic terms. A progressive sequence of assignments is designed to tackle the complexities of drawing the hand and arm:

- Studies of the skeleton of the arm and hand in constructional shapes and how these function (figs 110–113).
- Studies of the living hand investigating the elements involved in the construction and volume of the hand (figs 114–125).
- Finished studies of the hand (figs 126–131).
- Studies emphasizing spatial aspects (figs 130, 132).
- Freely invented hand gestures (figs 133–139).
- Free studies of the hand emphasizing expression (figs 140–143)

These are followed by studies of the living arm as a whole, concentrating on understanding the constructional form of actions by the arm and hand (figs 144–151).
Coordinating a sequence such as this reveals how close observation is called for to start with, while in the end what is required is a quickly executed study.

What graphic criteria should be arrived at in dealing with the skeleton of the lower arm and hand (figs 110–113)?

- The first essential is to ascertain the direction in which the individual sections of this part of the body run.
- Then work out the constructional shapes of all joints, proceeding from the grooved transverse roller of the elbow joint by way of the ovoid wrist joint and the ball-and-socket joints at the base of the fingers (see especially figs 111–113) on to the hinge joints at the ends of the fingers.
- Grasp the complexity of the way the radius and the ulna twist round one another (figs 110 center, 112 left) and of the metacarpus, paying special attention in the latter to the gradient of the arch of the back of the hand (fig. 112).
- Pay particular attention to the formal connection of the radius, via the inside of the metacarpus, to the index finger (esp. fig. 113).

In constructing a graphic study of the living hand the following points apply:

- Each individual finger has its own spatial position within the overall curve of the bone structure of the hand (esp. figs 114, 116, 117, 132).
- The convexities of the masses formed by the soft forms (such as the palm) meet, so creating mounds and hollows.
- Emphasize the change from soft to bony forms (esp. figs 118, 126, 129, 131).
- Always remember: every position of the hand is functional – lying, hanging, clenched or curled up, gripping, forming a cup-shape, etc.
- Use tapered shapes for the palm and fingers (figs 15, 116, 120–123, 125, 132) otherwise you will end up with sausages.

In finished studies of the hand, the main criteria are these:

- Pay careful attention to the most delicate intersections right down to the last finger tip.
- Work out the finer structural details such as knuckles, muscles and joint reinforcements (esp. figs 124, 126, 129, 130, 136), following the forms of the large and small curves with hatching.

The following procedure is recommended for studies of hand gestures:

- Using very simple methods and not getting bogged down in anatomical detail, look for the most attractive, eloquent form of the hand, imagination. Work very quickly (figs 133–135, 137–139).
- For this exercise use a medium that can keep pace with your imagination.

For free studies emphasizing expressiveness these factors are especially important:

- Heightening the functional expression (figs 119, 132, 140–142).
- Experimentation with different methods of working such as free linear recording (fig. 141) or well calculated coordination of masses and developed forms conveyed with a half-dry brush (fig. 143).

Studies of the living arm and attached hand concentrate on:

- Making constant basic forms and directions and the volumes of muscles layered over them visible (figs 144, 146, 147, 148).
- Articulating the functions and platic behavior of the arm convincingly as a preparation for the specific way in which the hand is joined (figs 145, 146, 148).
- Working out the main volumes and their changing positions in space (figs 144, 146, 148).

With an increasingly secure grasp of the construction, function and plastic behavior of the hand and arm the student can move on to freer procedures. A relaxed approach may treat the soft, curving forms of skin and musculature in an open web of lines and create contrasts with the firm, stable bone shapes (figs 149, 150, 151). Once you are sure of your skill and know your subject, you can use an artistic approach to life studies, and be free to omit, abbreviate, use symbols, concentrate and condense, use abstraction and combination; a freedom that will enable you to cope both with diversity of form and what can be seen at any given time – particularly when dealing with the arm and the hand.

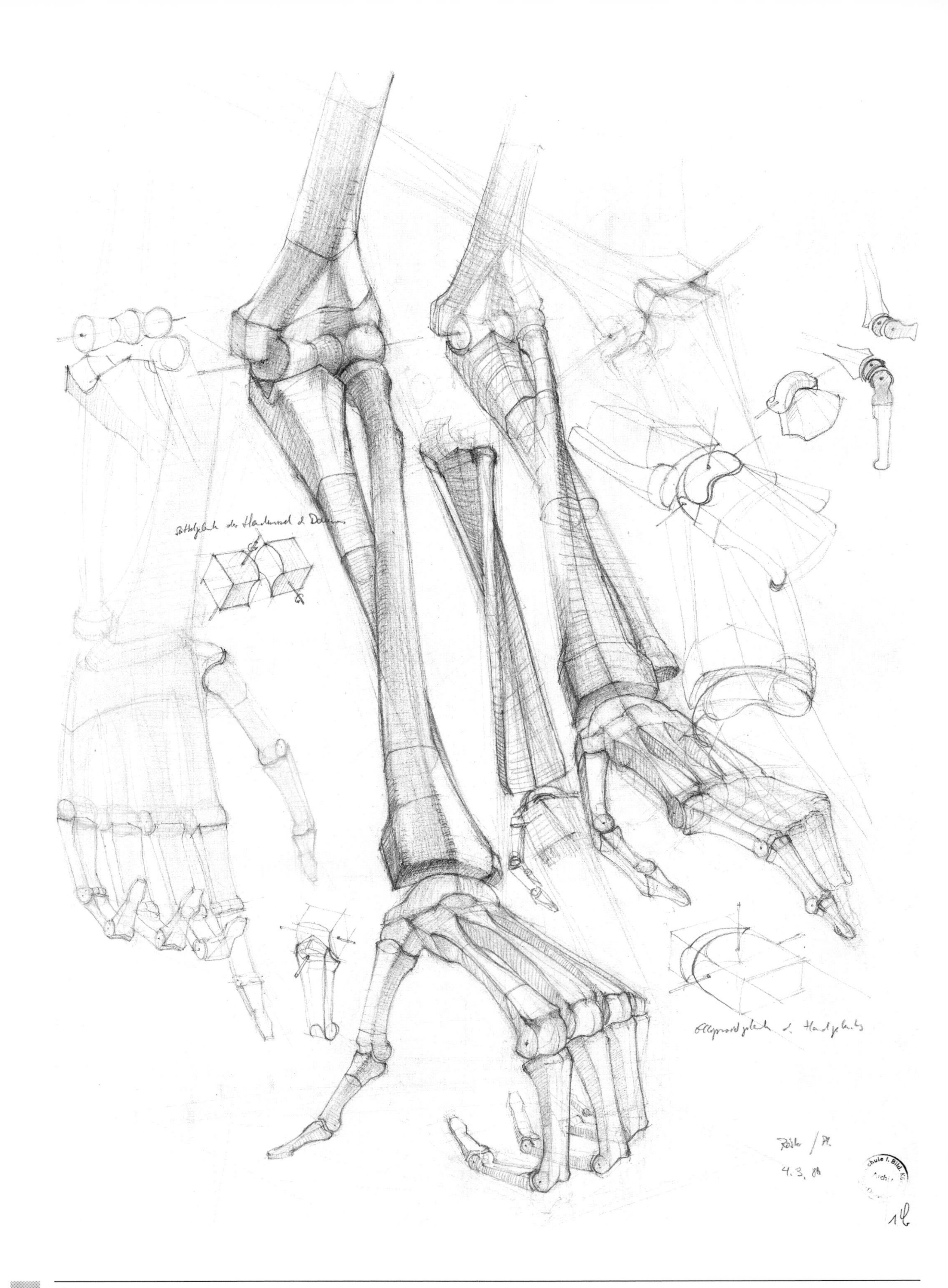

110 HOW A VARIETY OF DIFFERENTLY SHAPED JOINTS FIT INTO THE SKELETON OF THE LOWER ARM

The differently shaped joint forms, from the complex elbow joint by way of the ovoid wrist joint, the saddle joint of the thumb or the hinge joints of the fingers, are brought together in the main drawing from a number of subsidiary studies, and fitted into the overall constructional context.
Student of sculpture, fourth semester

111 A GRASP OF COMPLEXITY AND DEPICTION IN DETAIL WORKING TOGETHER

In the upper section of the drawing the back of the metacarpus is depicted as a coordinated curve, and in the lower part the details of the components of the arched shape are worked out. Constructionally, every single bone in the metacarpus is like a tension bridge. All drawings of joints – even when they are looked at constructionally – tell us something about their functional capacity.
Student of sculpture, fourth semester

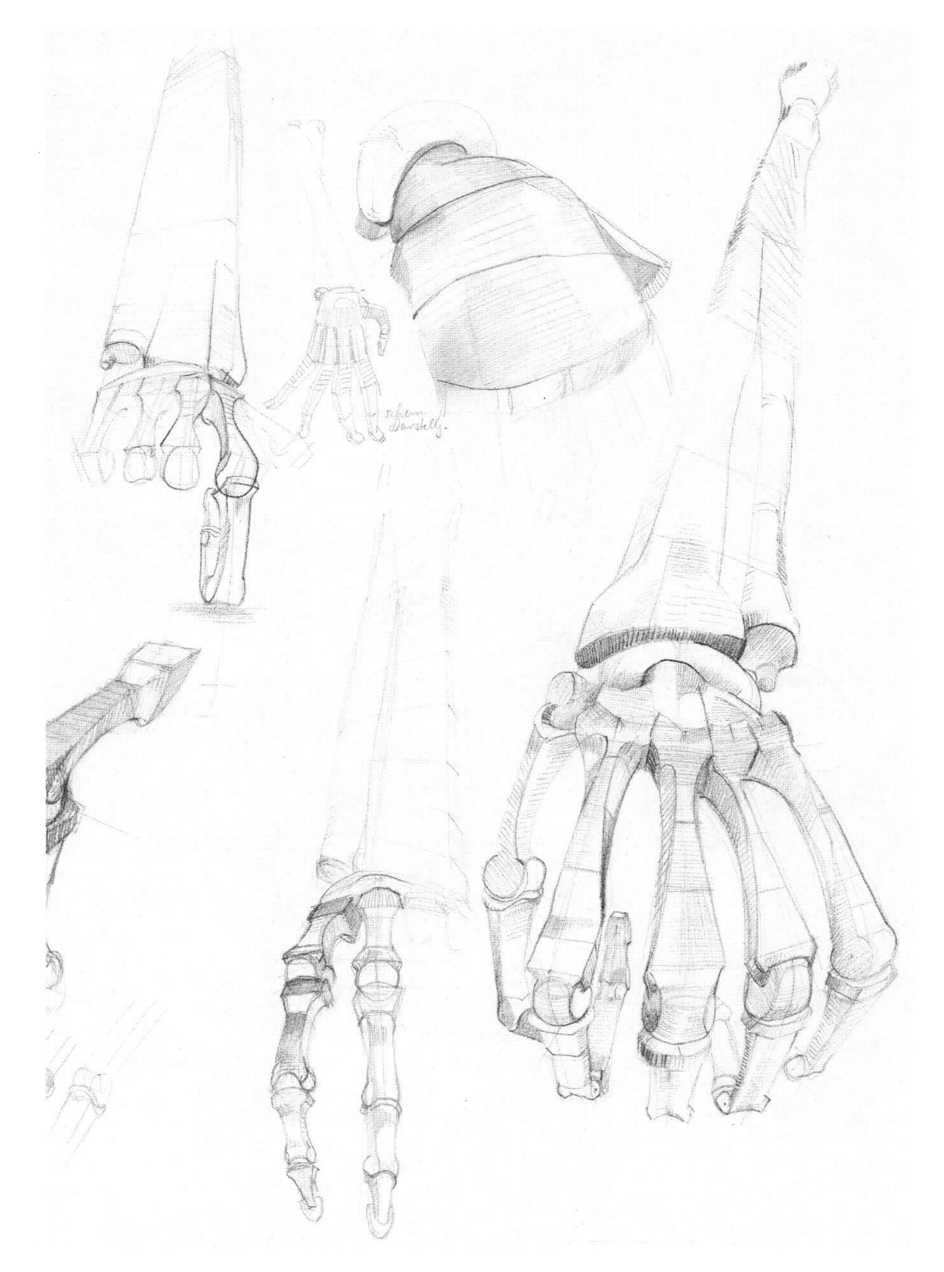

112 REINFORCING SOLIDITY BY INDICATING THE SPATIAL DIRECTIONS OF THE GRADIENTS

Here, the hatching follows the curves of the arch on the back of the hand; this is repeated further down the articulation, and forms the basis of the position of each individual finger.

Student of painting/graphic art, fourth semester

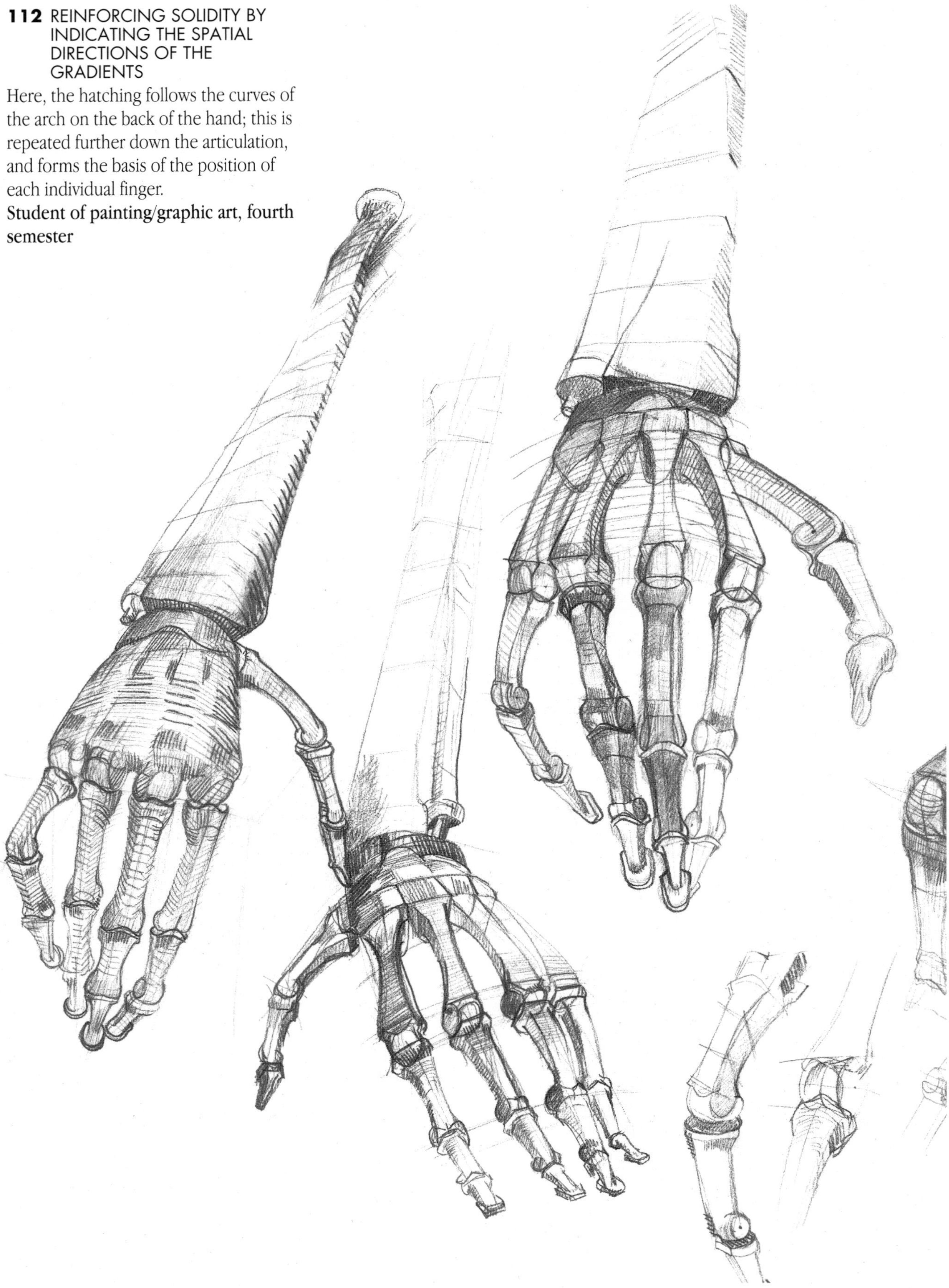

113 SHORTHAND VERSION OF THE SKELETON OF THE LOWER ARM AND THE HAND

As soon as the constituent parts of the structure and functional essentials have been recognized and learned, the methods used in depicting the hand are almost automatically simplified, right down to the shorthand of a purely linear definition.

Student of restoration work, fourth semester

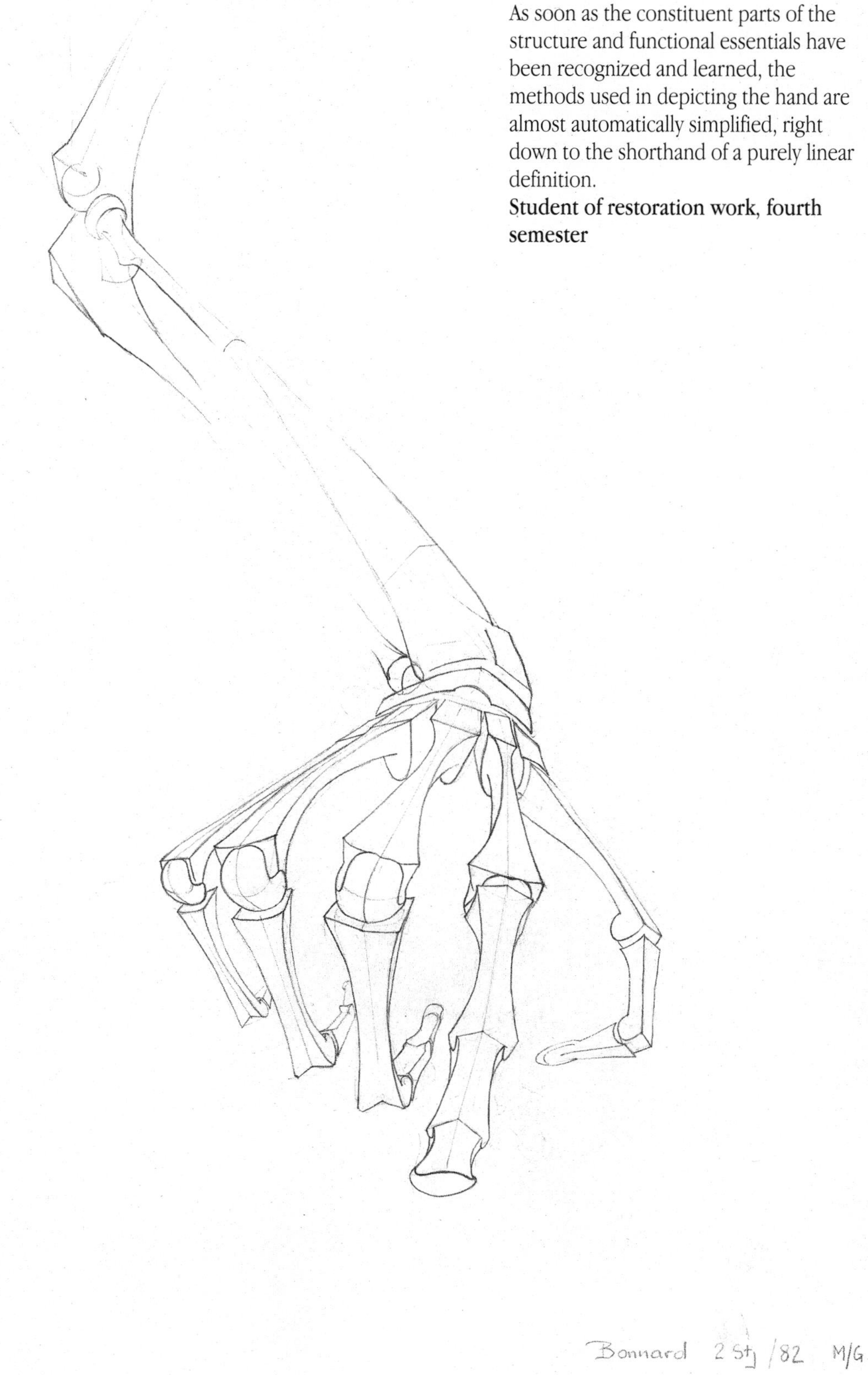

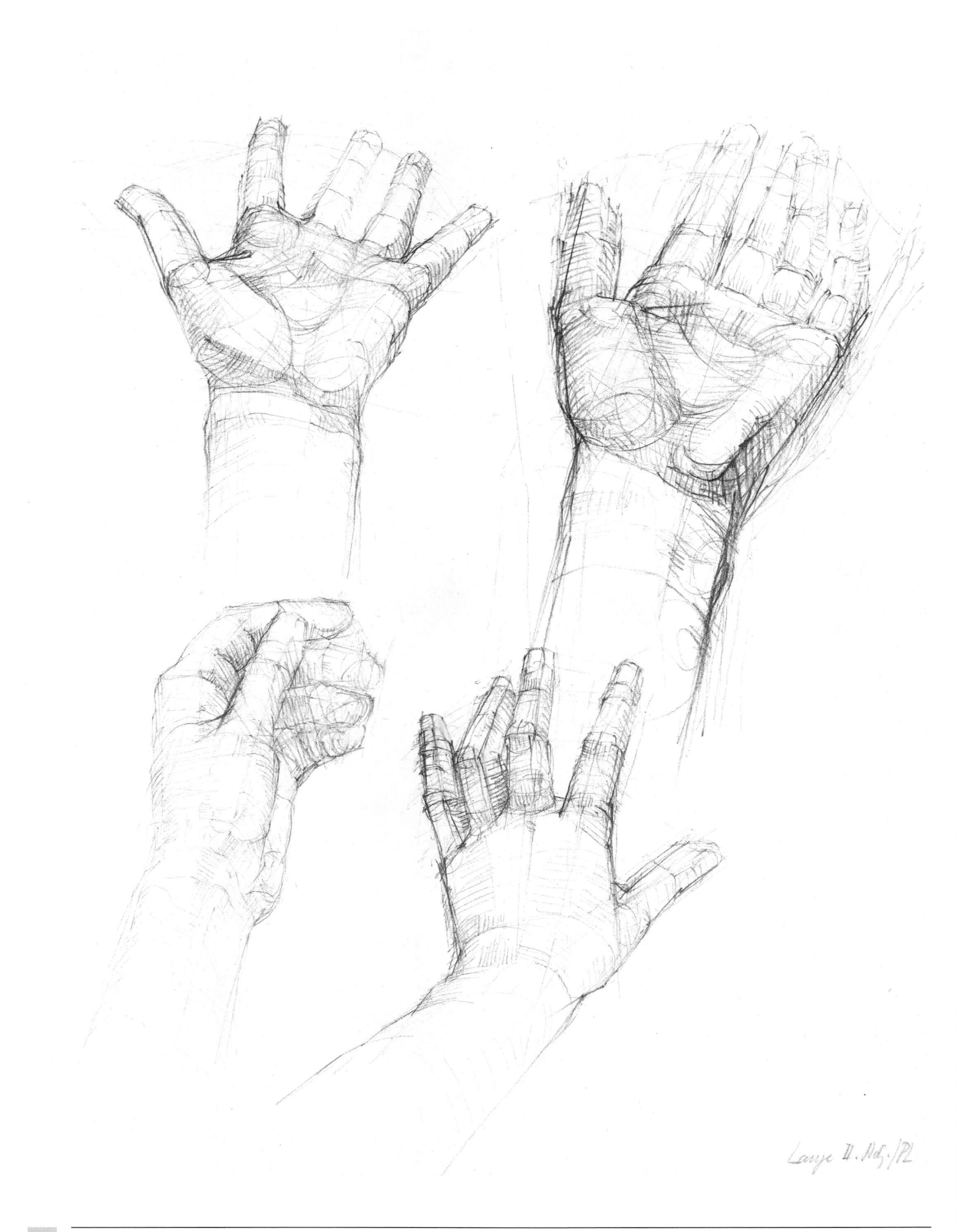

Chapter 8

Studies of the construction, functioning and plastic behavior of the hand and arm

114 A STRUCTURAL APPROACH COMBINED WITH DIRECTIONAL DEFINITION AND MASSING OF VOLUMES

The way the hand is built up on the basis of clearly arranged volumes makes it easier to control the problems of foreshortening. There are also clear directional lines and continued articulations which give a sense of stability.

Student of sculpture, fourth semester

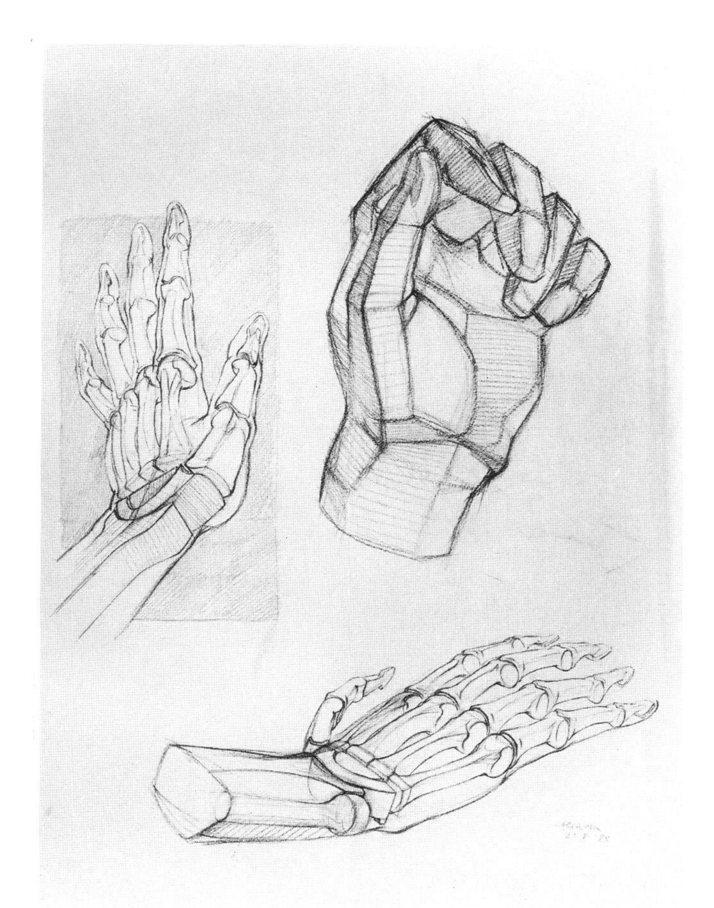

115 THE FRAMEWORK OF THE HAND WITH BLOCK SHAPES

Problems of foreshortening are easier to master if the marked differentiations of form are first converted into simple block shapes.

Student of set painting, fourth semester

116 THE UNITY OF OVERALL AND INDIVIDUAL CURVES IN A STUDY OF THE HAND

The main task was to convey this unity, drawing in only the curves relating to a spatial view. Students were allowed to draw in a minimal number of stabilizing contour lines only once this had been done.

Student of painting/graphic art, fourth semester

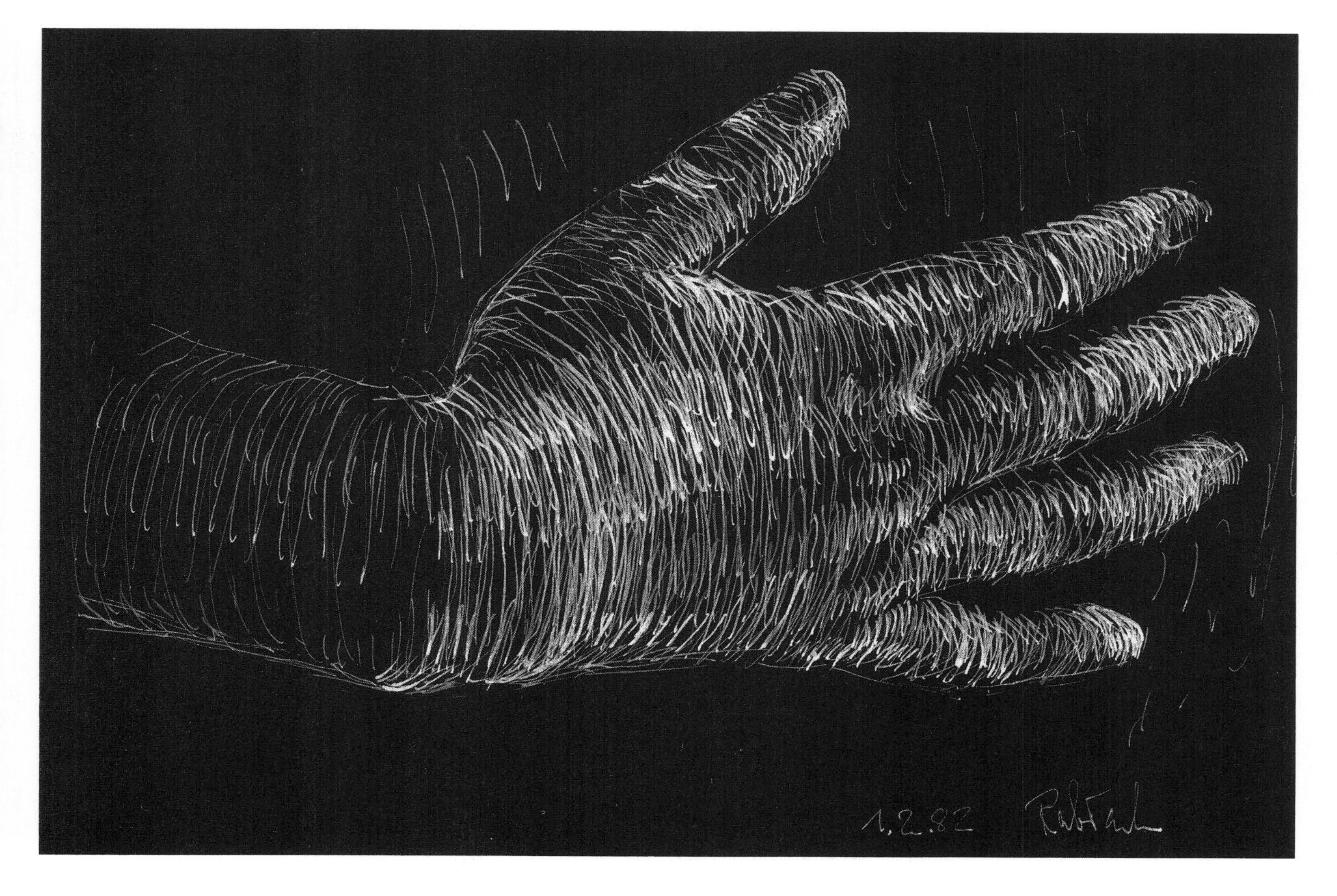

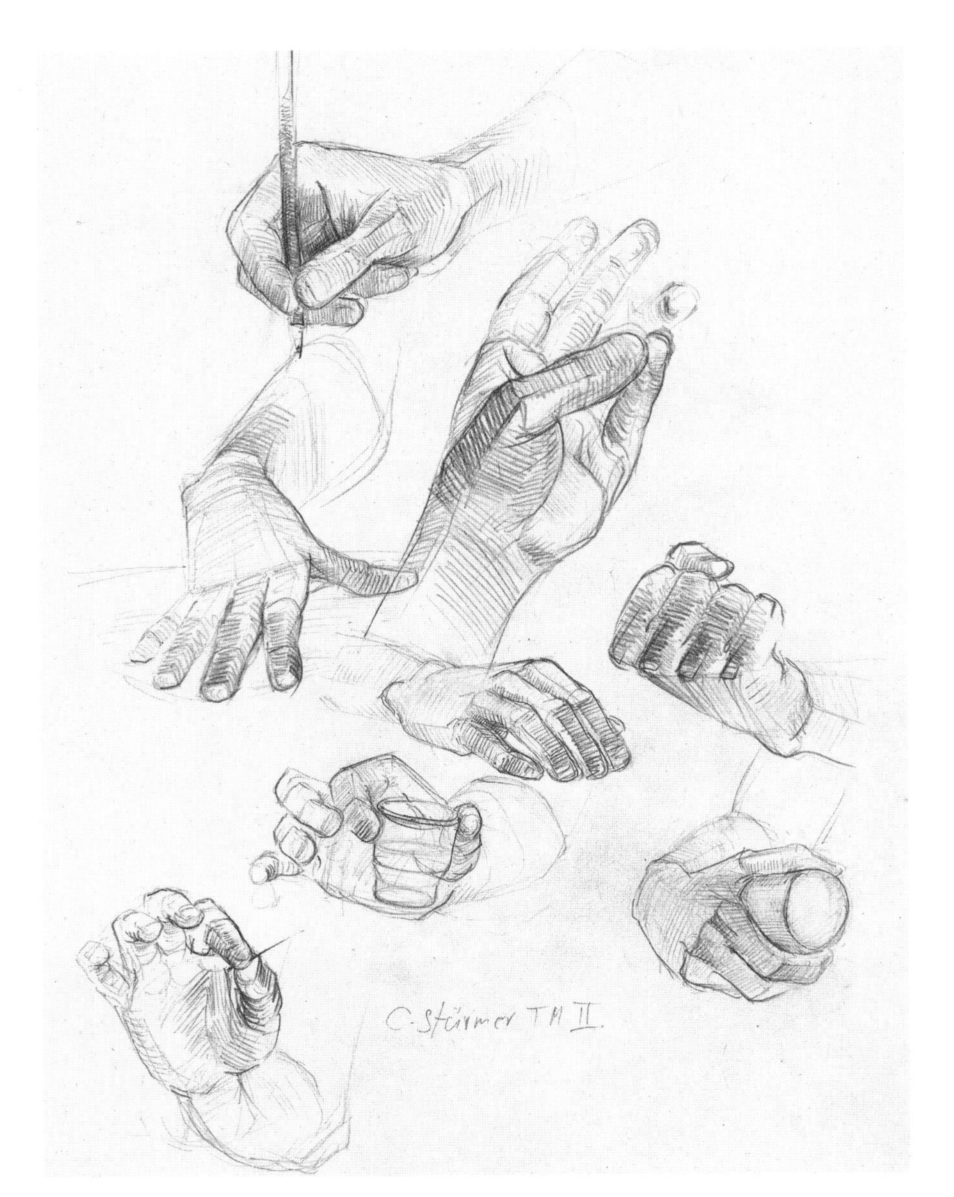

117 SYNTHESIS OF FUNCTION AND PLASTIC ORGANIZATION IN THE HAND

The scope of the hand in action ranges from the clenched fist to the holding of delicate objects. The clearly organized positions of the lines denoting solidity also serve here to clarify the functional processes.

Student of set painting, fourth semester

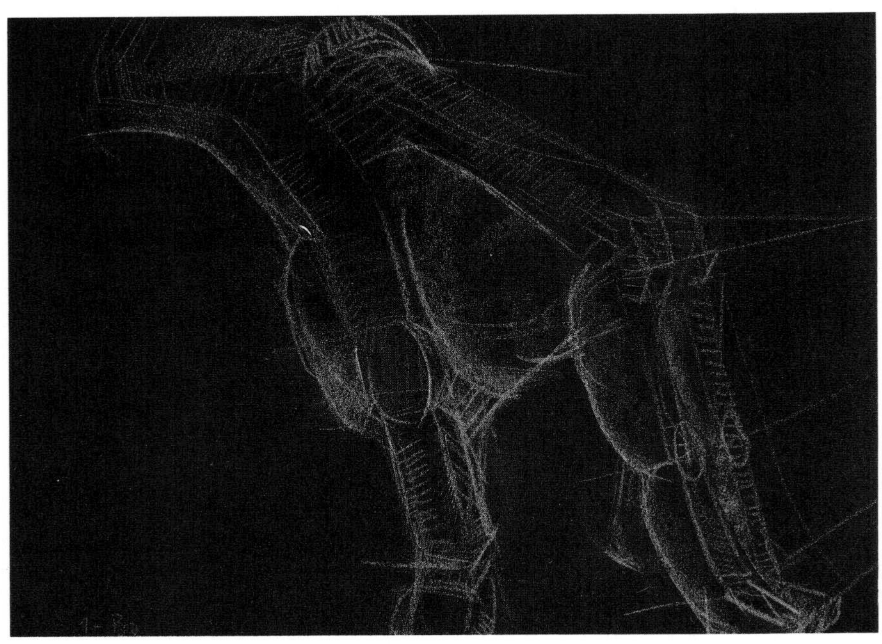

118 STUDY OF THE HAND CONCENTRATING ON STRUCTURE

The rigid forms of the skeleton are in marked contrast to the padded soft forms.

Student of painting/graphic art, fourth semester

119 SPATIAL AND FUNCTIONAL EXAGGERATION

This study, largely based on visualization, expresses the forward thrust of the hand by exaggerating its size through the progressions and by the marked articulation of the finger positions.

Student of painting/graphic art, fourth semester

120 CONVEYING SOLIDITY WITHOUT USING OUTLINES TO DEFINE SPATIAL TRANSITIONS

The study is based on indicating the different spatial gradients of the planes and marking the larger joints; no contour lines are used in the drawing and this produces fluid spatial transitions.
Student of painting/graphic art, fourth semester

121 TWO HANDS PLACED TOGETHER

The underlying purpose of the study is to merge two separate bodies into a single visual unit. It is also important to clarify the way the shape of the upper hand fits over that of the lower one.
Student of painting/graphic art, fourth semester

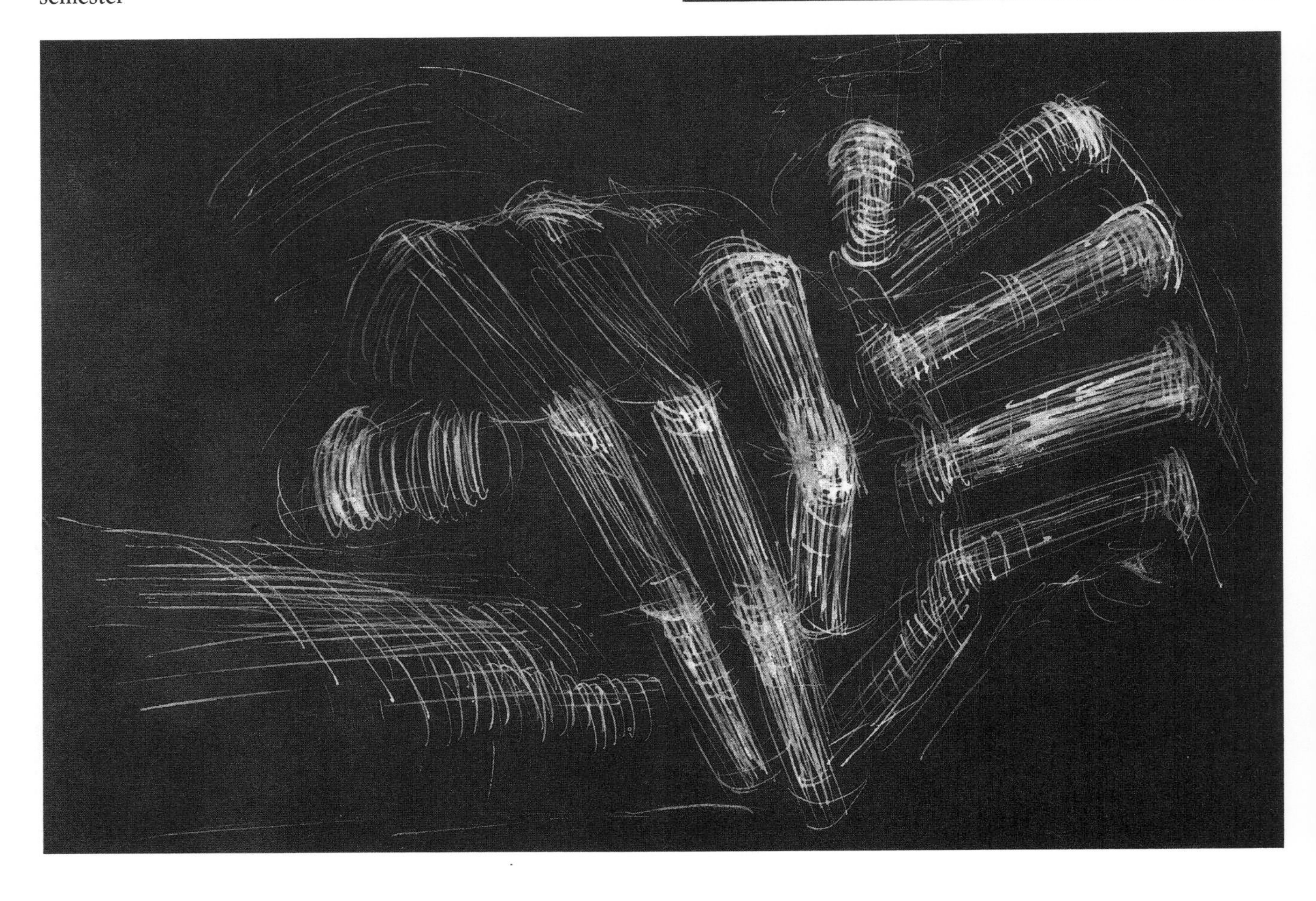

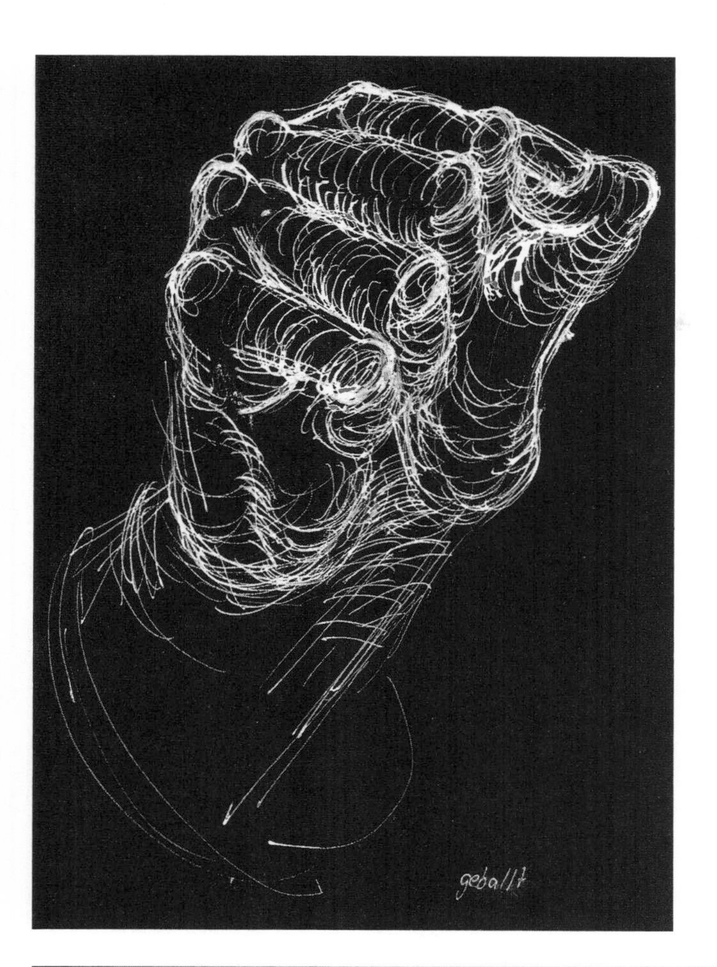

122 ORGANIZING OVERALL MASS AND INDIVIDUAL SHAPES

Individual volumes are embedded into the mass of the hand which is clenched into a fist while the bent finger joints are treated like pieces of angled pipe.
Student of painting/graphic art, fourth semester

123 CHOOSING APPROPRIATE GRAPHIC TREATMENT FOR THE FLOWING FORMS OF THE BACK OF THE HAND AND FINGERS

Very few contour lines have been used, and an attempt has been made to emphasize the effect of the smoothly flowing shapes with parallel hatching.
Student of painting/graphic art, fourth semester

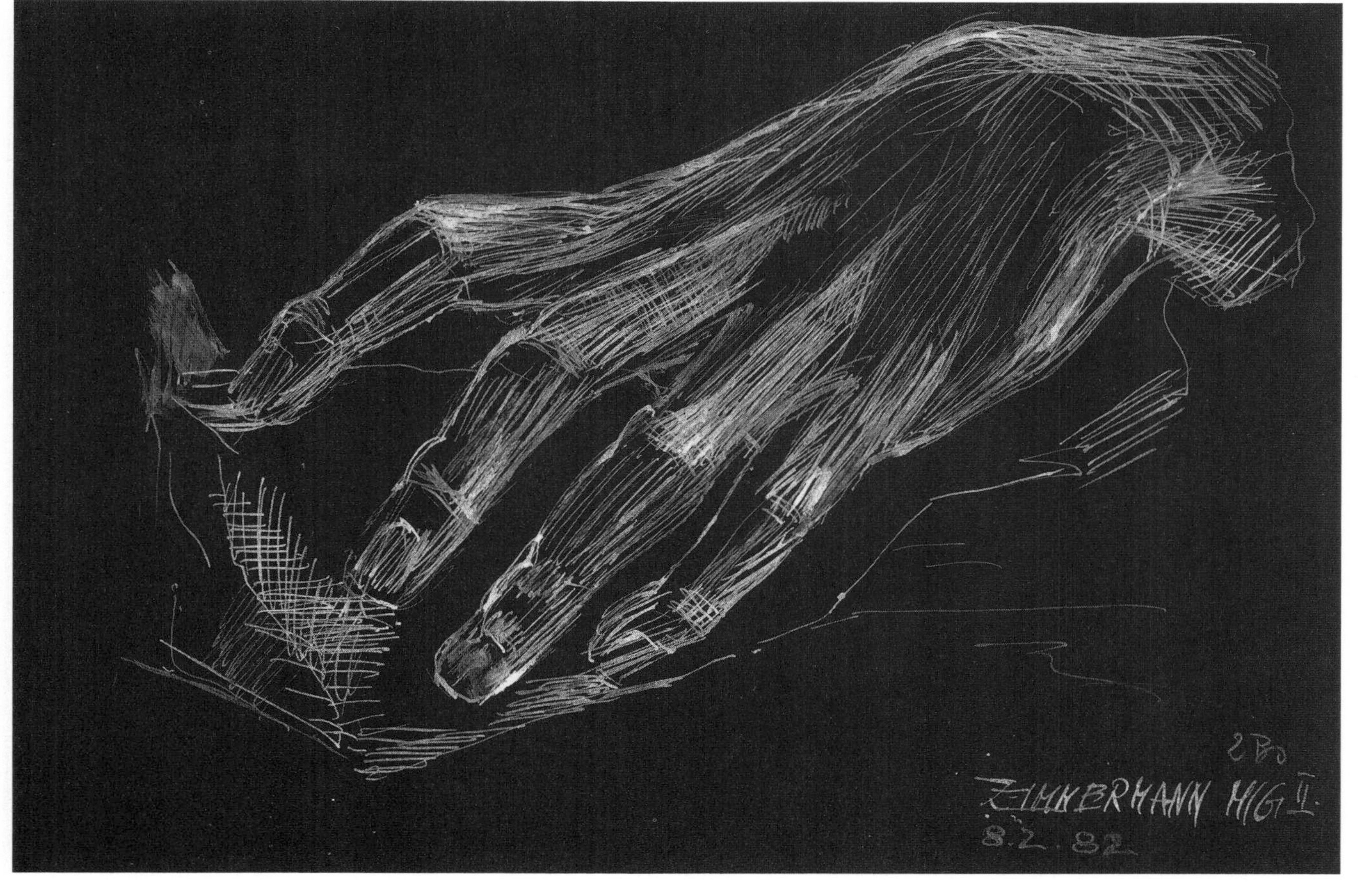

124 HIGHLIGHTING AS A MEANS OF DEVELOPING FORM

The use of white chalk on a black ground gives gradations of lightness which both highlight forms and enhance or understate spatial factors.
Student of restoration work, fourth semester

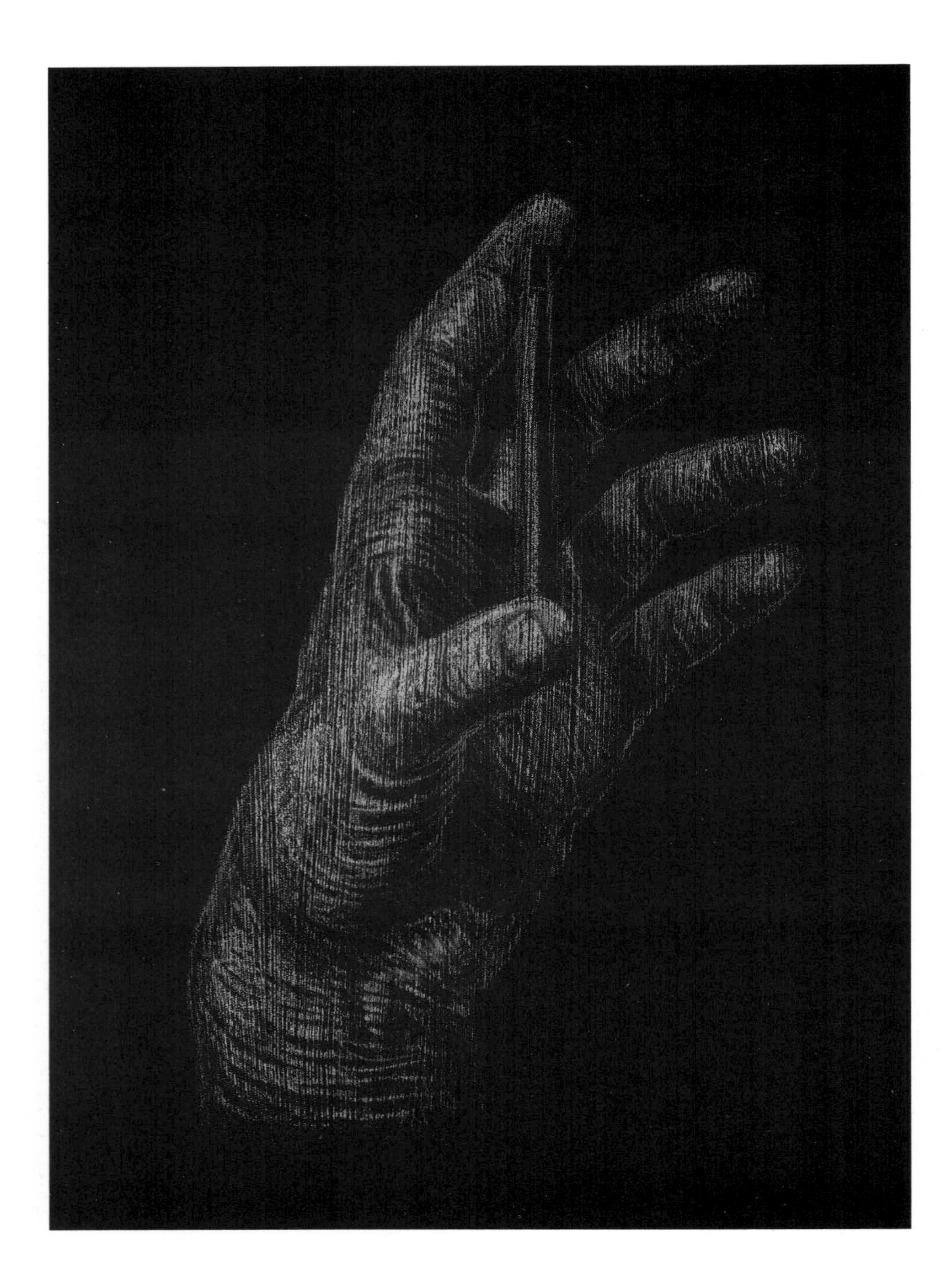

125 THE USE OF SHARP EDGES
The powerful impact of the outline shape is produced by sharply turning contour lines on the one hand, and close cross-hatching within the internal shapes enclosed by hard facets on the other.
Student of painting/graphic art, fourth semester

126 FINISHED STUDY OF THE HAND
Several aspects are realized using very sensitive hatching executed with a pointed nib: perspective depth, foreshortening and intersections, articulation of the shapes of the fingers emerging from the volume of the palm, the distinction between bony and soft parts, and the spatial recesses between convex volumes.
Student of painting/graphic art, fourth semester

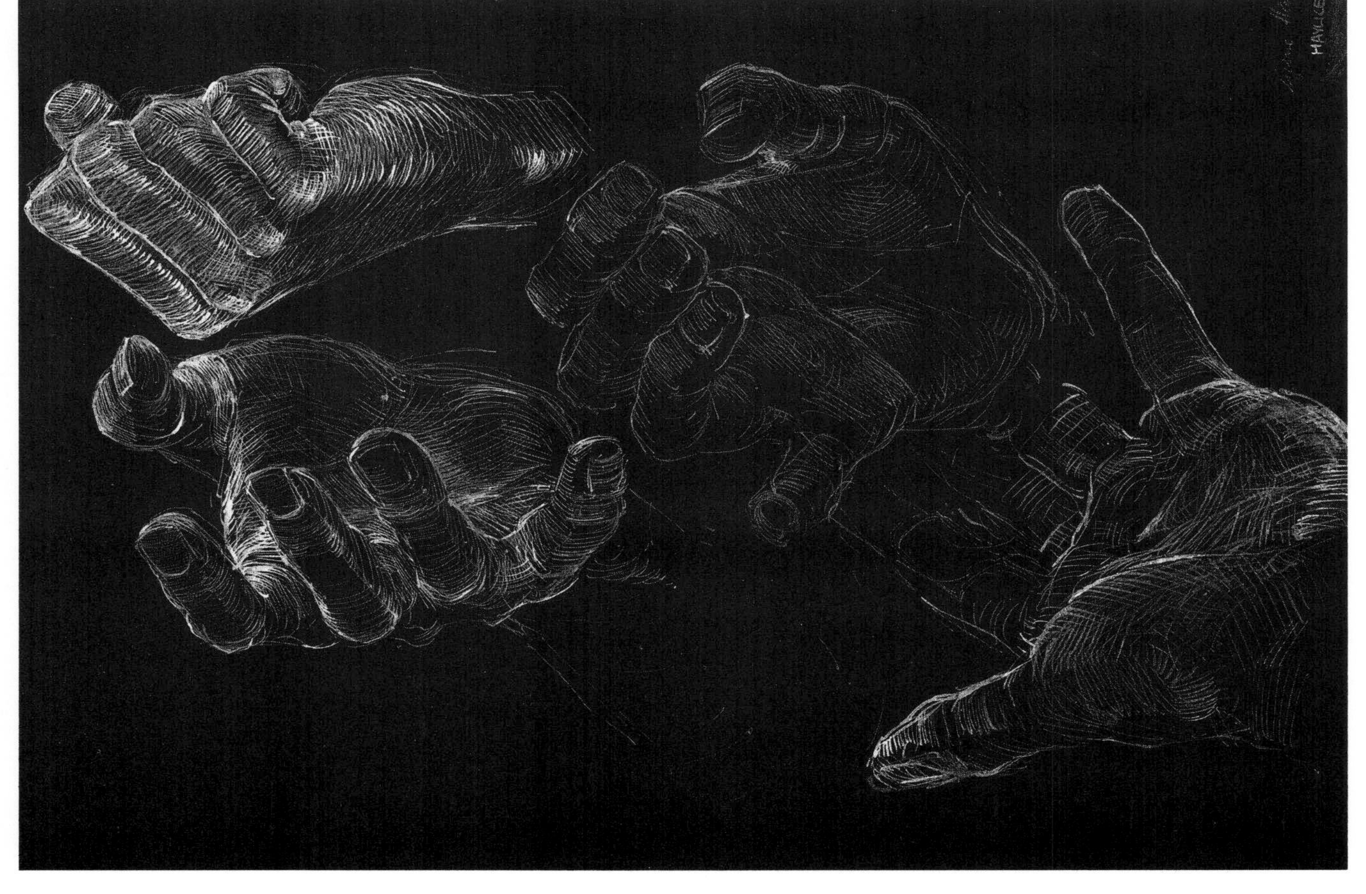

127 CONTRIBUTION MADE TO FUNCTIONAL PROCESSES BY DEPICTING CREASES OR FOLDS

Placing the thumb against the finger tips is always accompanied by compression of the skin between the thumb and the index finger. Indicating this graphically also emphasizes the functional process.

Student of set painting, fourth semester

128 UNITY AND SIMPLICITY OF FORM

As well as differentiating forms when making graphic studies of the hand, you may be asked to create unified, simple forms, especially when considering basic functions of the hand, here depicted in a cupped and a hooked shape.

Student of stage design, fourth semester

129 THE HAND AS A FORM OF 'SELF-PORTRAIT'

Students were asked to look at their own hands for those characteristics which may be expressed graphically, and seen as an expression of their personality and appearance as an overall formal statement.

Student of painting/graphic art, fourth semester

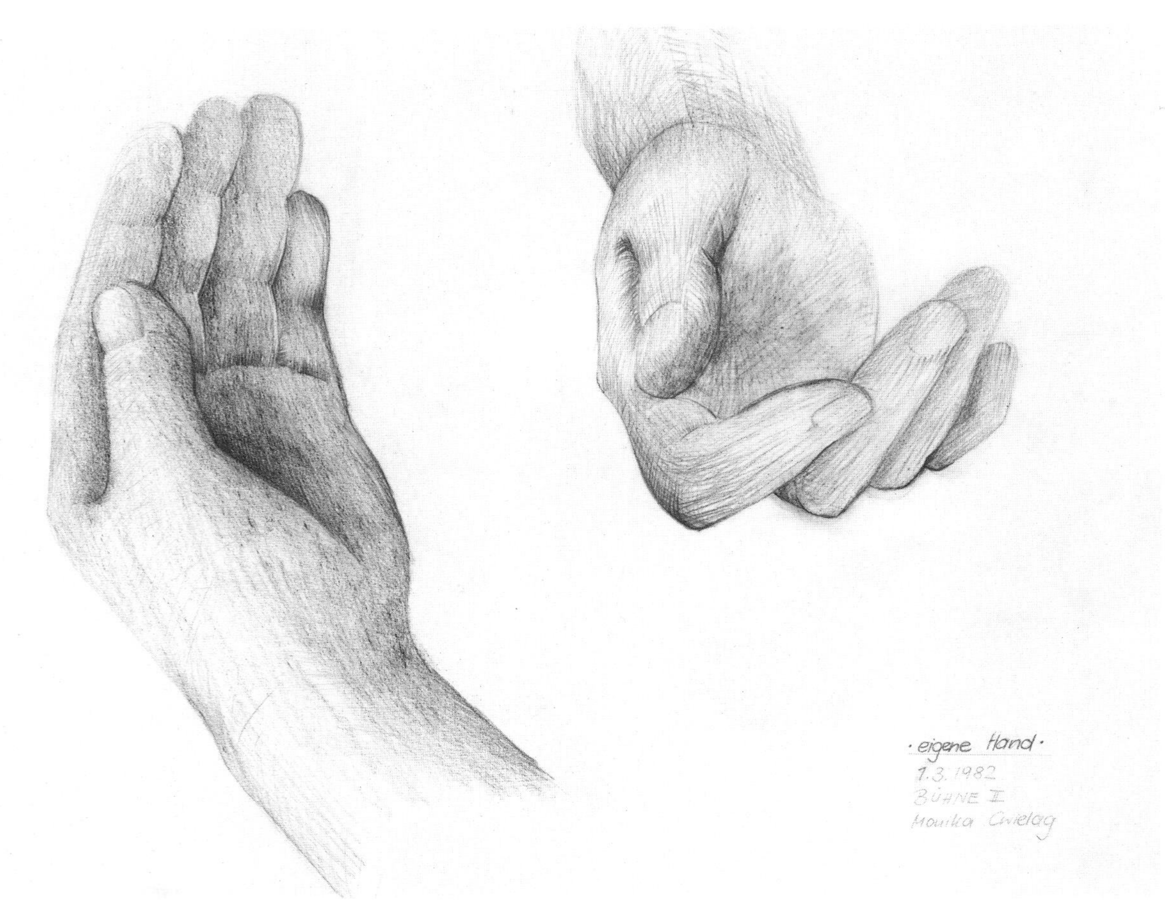

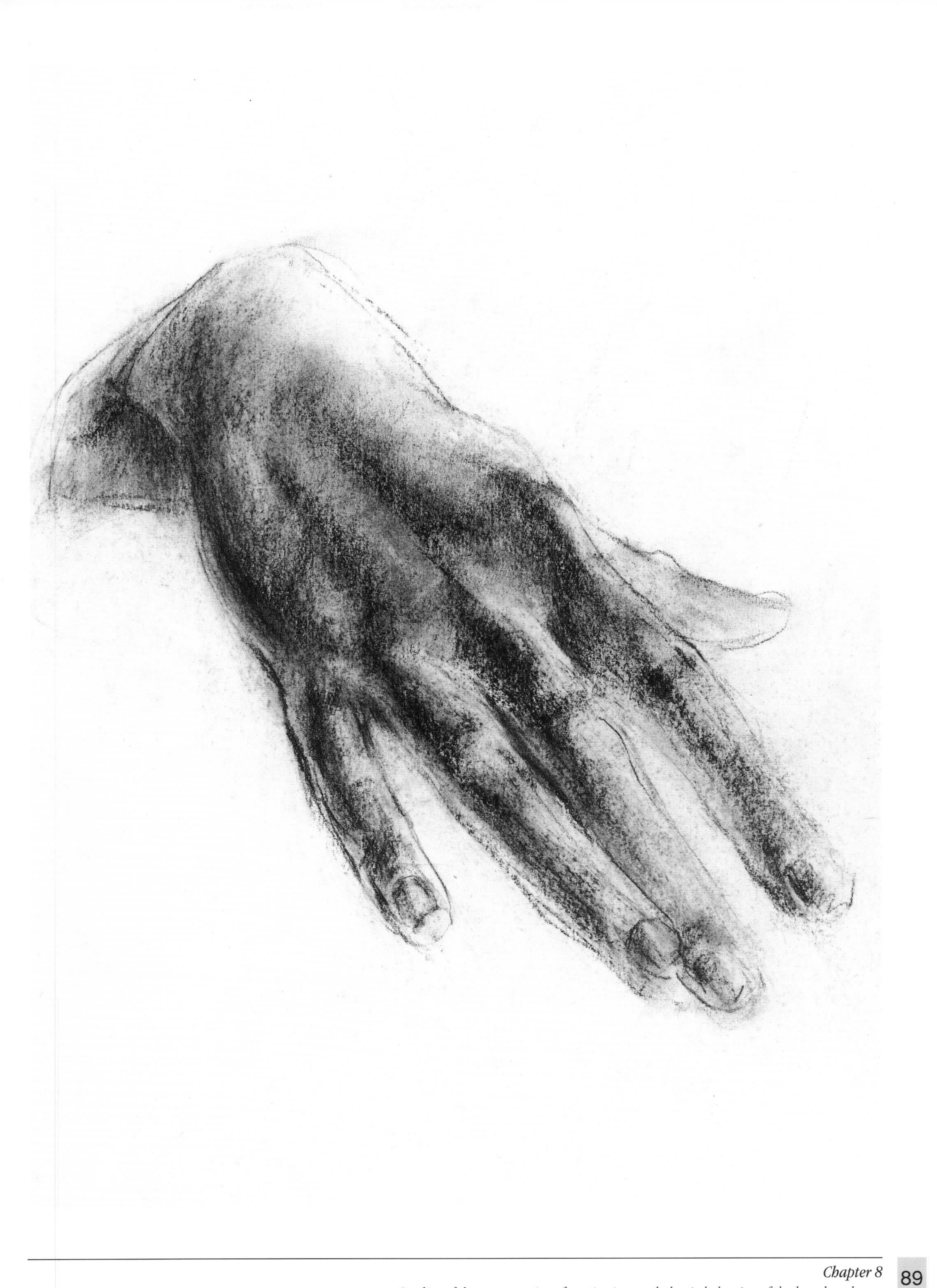

130 THE HAND AS A CREATOR OF SPACE

As well as mastery of anatomical and functional factors, it is a vital and equally important part of your study to take account of the spaces enclosed by the hand.

Student of painting/graphic art, fourth semester

131 SURFACE TEXTURE

In finished studies of the hand, the experiential quality of surface textures can be highlighted, so enhancing their expressive power: when the hand is tight shut the knuckles and muscles stand out, with the skin tightly stretched across them like thin parchment.

Student of painting/graphic art, fourth semester

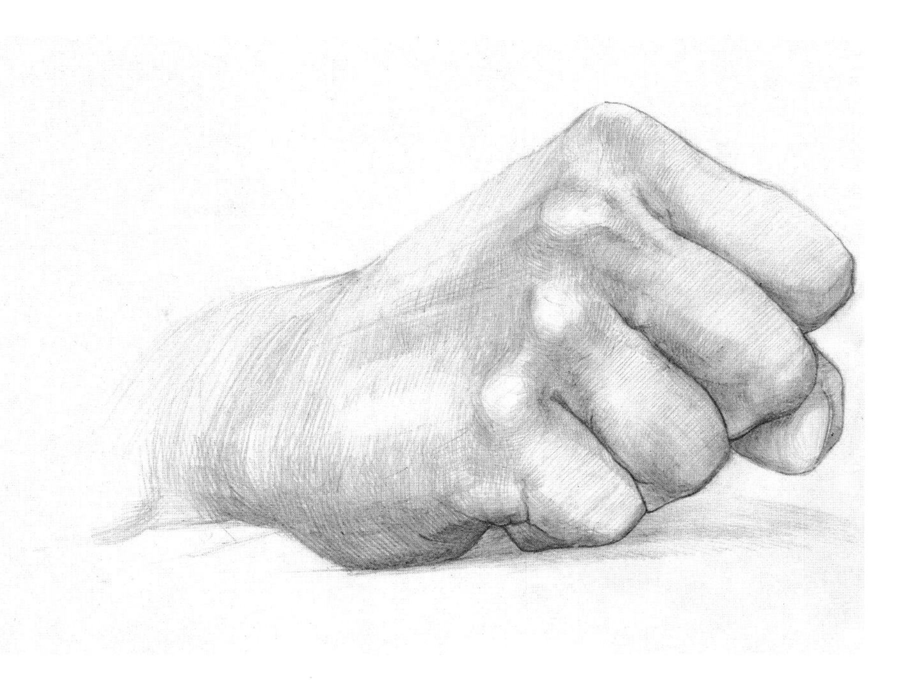

132 THE RECIPROCAL INFLUENCE OF THREE-DIMENSIONAL AND SOLID REPRESENTATION

The set task: draw only the spatial gradients of the fingers and the metacarpus, careful weighting nearest and farthest parts by hatching in differing densities. Result: by investigating spatial factors you convey solidity.

Student of painting/graphic art, fourth semester

133 IN SEARCH OF HAND GESTURE

Basic hand gestures, here indicating that something should stop, can be more surely and easily found with some use of the imagination. Use a medium that can keep pace with your imaginative powers. Painstaking elaboration can sometimes hinder rather than promote the spontaneity of the concept.
Student of painting/graphic art, fourth semester

134 QUICKLY SKETCHED VISUALIZATION OF A HAND GESTURE

Trying out the most attractive imaginary view, in which the gesture has impact combined with spontaneous sketching, often expresses the symbolic value of the hand.
Student of painting/graphic art, fourth semester

135 NON FINITO AS AN EXPRESSIVE MEANS OF STUDYING HAND GESTURES

Parts that have been left open and unfinished on purpose, can work as indications of graphic spontaneity, turning the hand into an expressive instrument.
Student of painting/graphic art, fourth semester

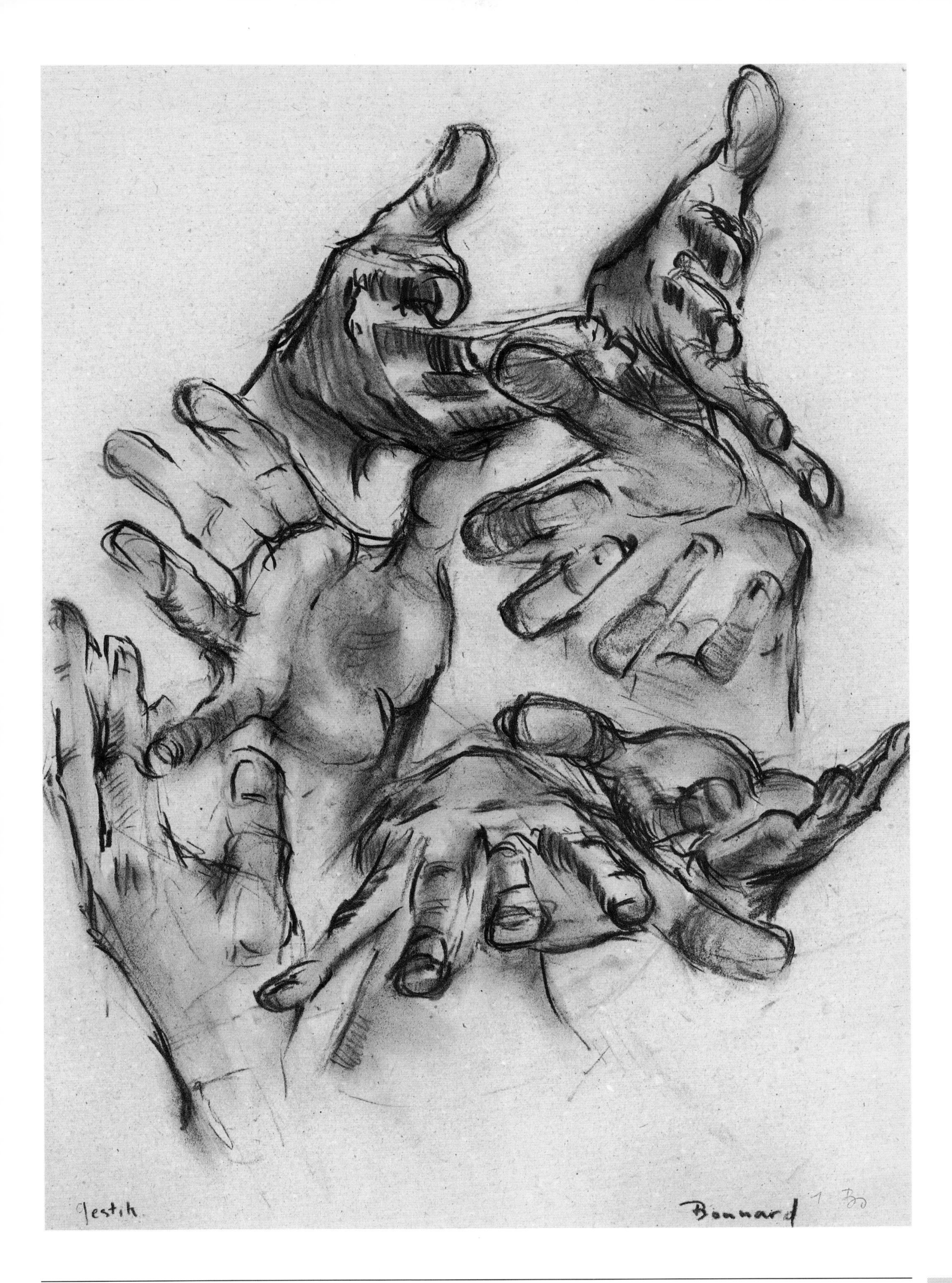
Gestik
Bonnard

136 A CAREFULLY CONSTRUCTED HAND PRODUCED AFTER SKETCHES OF HAND GESTURES

A free approach to the expressive drawing of gesture and finding the most impressive vocabulary for it, often using a fragmentary style, should be backed up from time to time by close studies.
Student of painting/graphic art, fourth semester

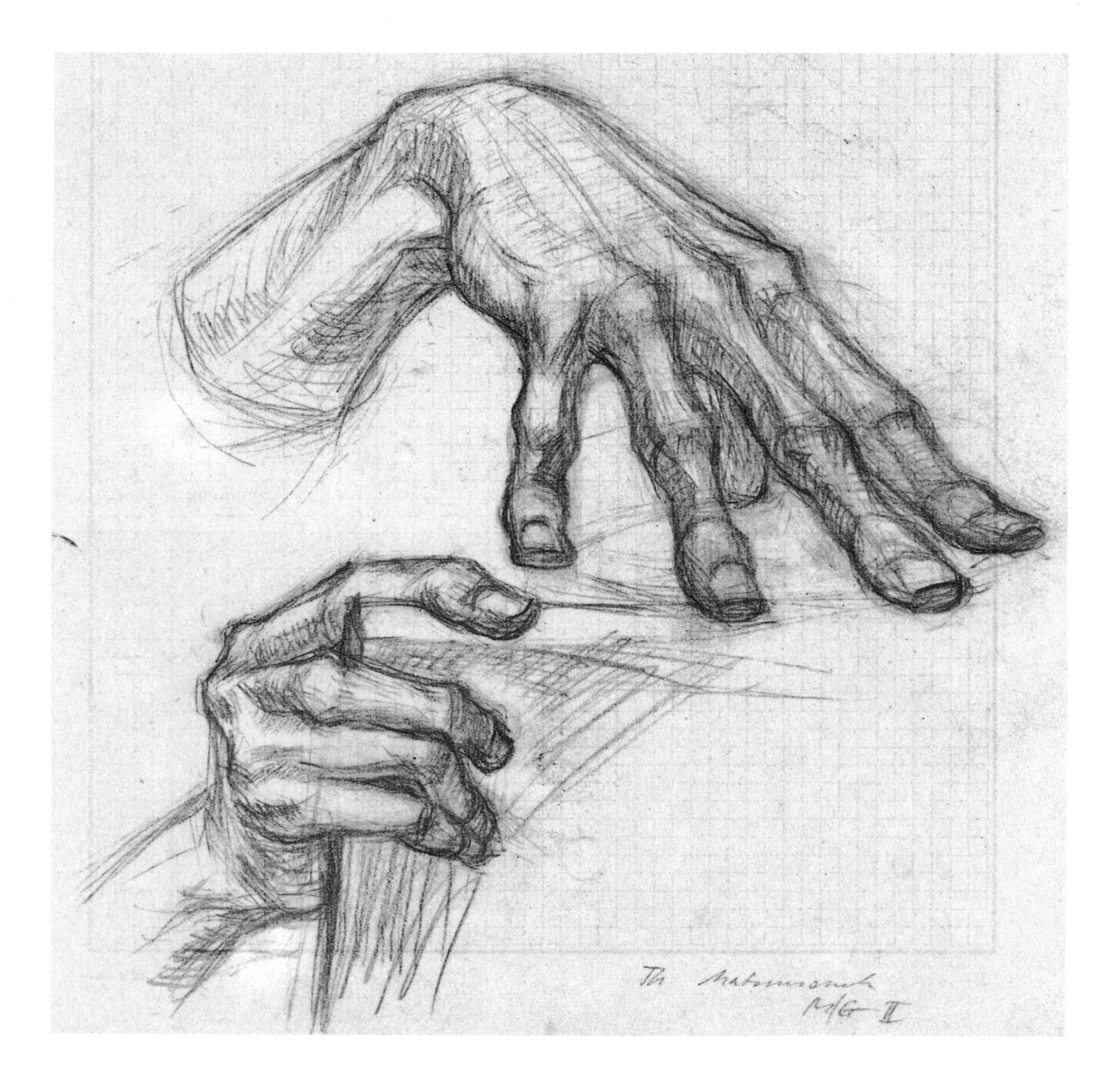

137 A RESPONSIVE MEDIUM FOR GRAPHIC INVENTION

To record ideas quickly when visualizing hand gestures, here the gesture of pointing, the means used must be spare.
Student of painting/graphic art, fourth semester

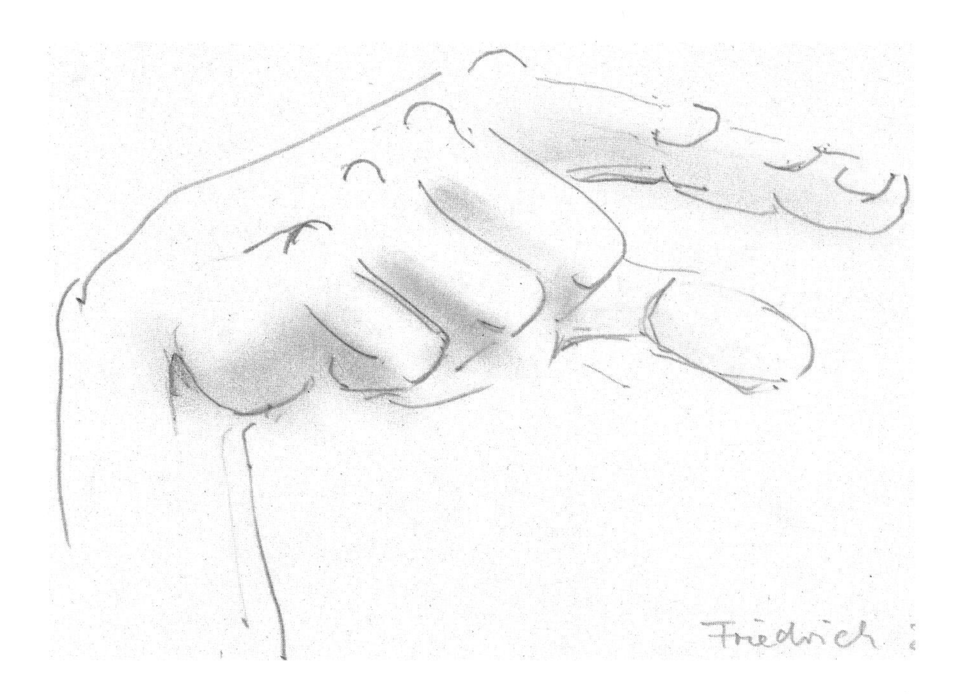

138 EXPLOITING THE ELOQUENCE OF THE BRUSH

The set task is the same as in fig. 137. Here straight, long brush strokes and light dabs are used together to convey a pointing hand.
Student of painting/graphic art, fourth semester

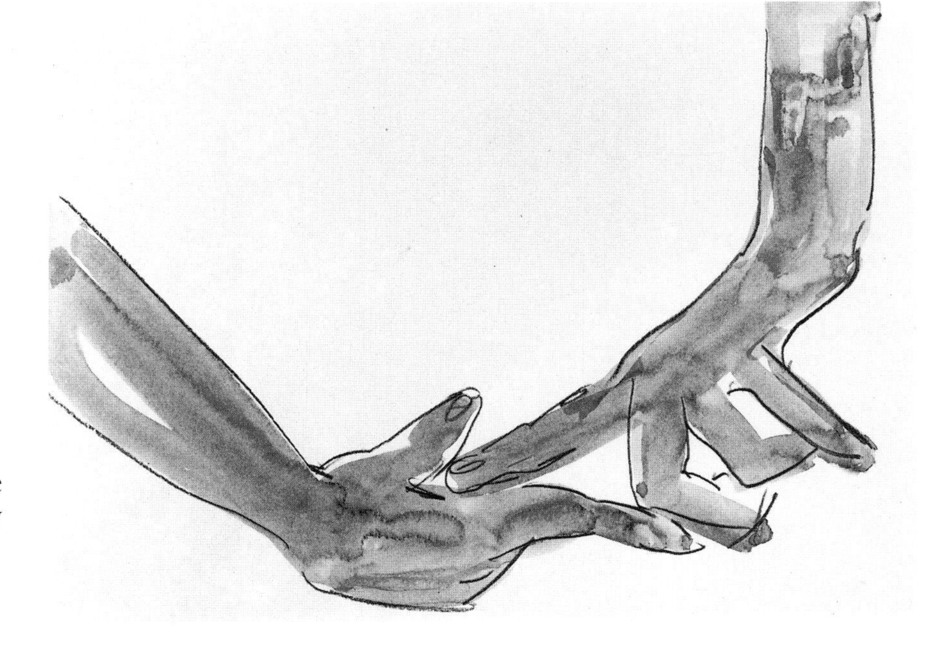

139 THE NARRATIVE HAND

A model with a gift for mime told a whole story using only hand gestures; this study records an instant from it.
From a Bammes course at the Schule für Gestaltung, Zurich

140 STUDY OF THE HAND WITH THE EMPHASIS ON EXPRESSION

The expressive character of the study relies on a few important components: the elongation of the proportions, the emphasis of the skeletal points and the functional exaggeration of the gestures.
Student of painting/graphic art, fourth semester

PAGES 98/99

142 HEIGHTENING THE IMPACT OF THE JOINTS

Excitement is demonstrated in this hand gesture by the exaggerated bending and twisting of the joints and the elongated sections of the fingers.
Student of painting/graphic art, fourth semester

143 EMPHASIZING THE UNITY OF THE MASSES

If we concentrate on amalgamating details into an assembled mass, there is no need for detail. The expressive power derives from the contrast of broad, open areas of black and delicate linear touches.
Student of painting/graphic art, fourth semester

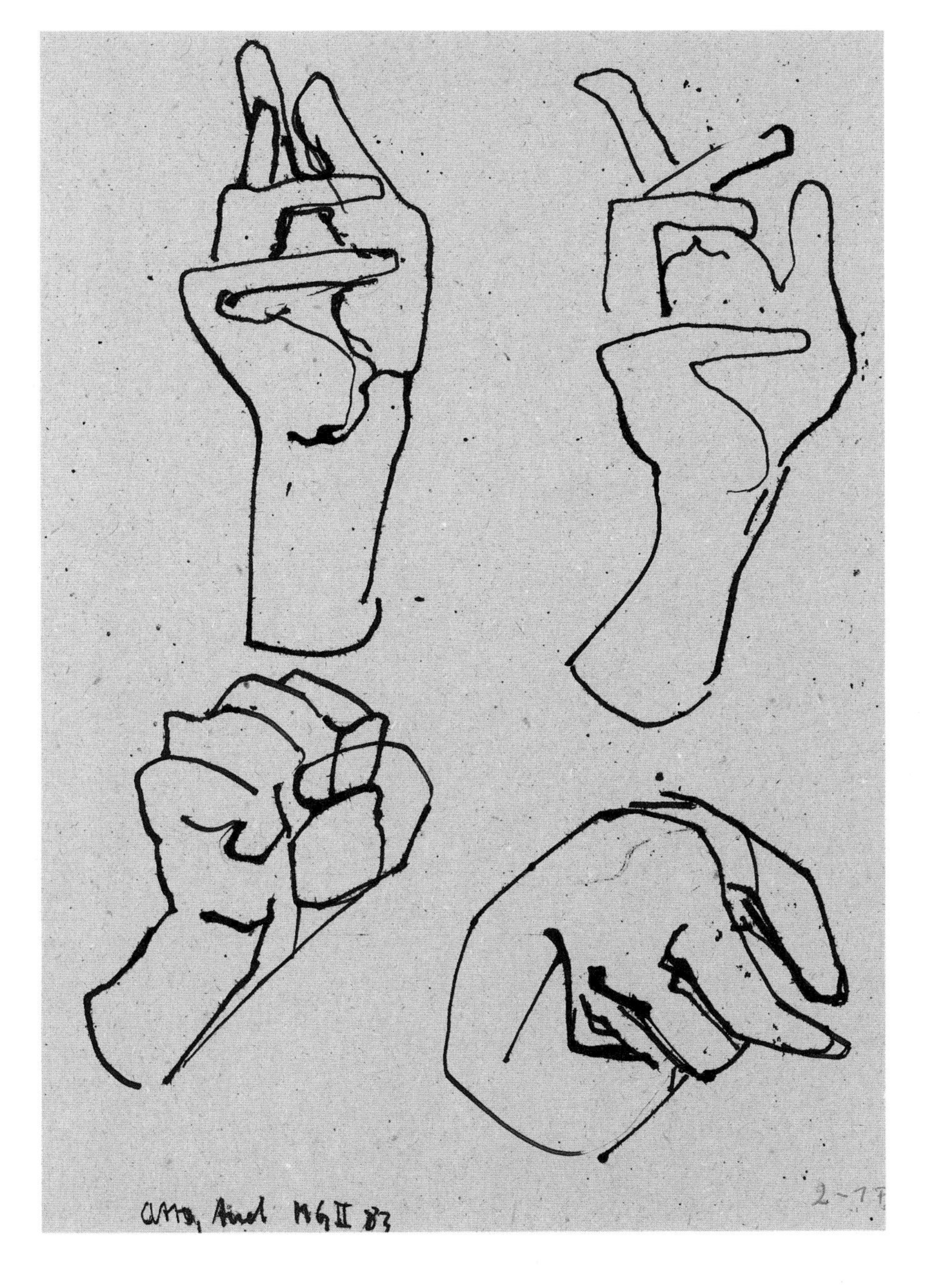

141 LINEARLY CONVEYED EXPRESSIVENESS OF THE HAND

In contrast to fig. 140 where the expressive content of the study relies largely on anatomical structure, this study is restricted to extremely economic linearity to convey the expression of the gesture.
Student of painting/graphic art, fourth semester

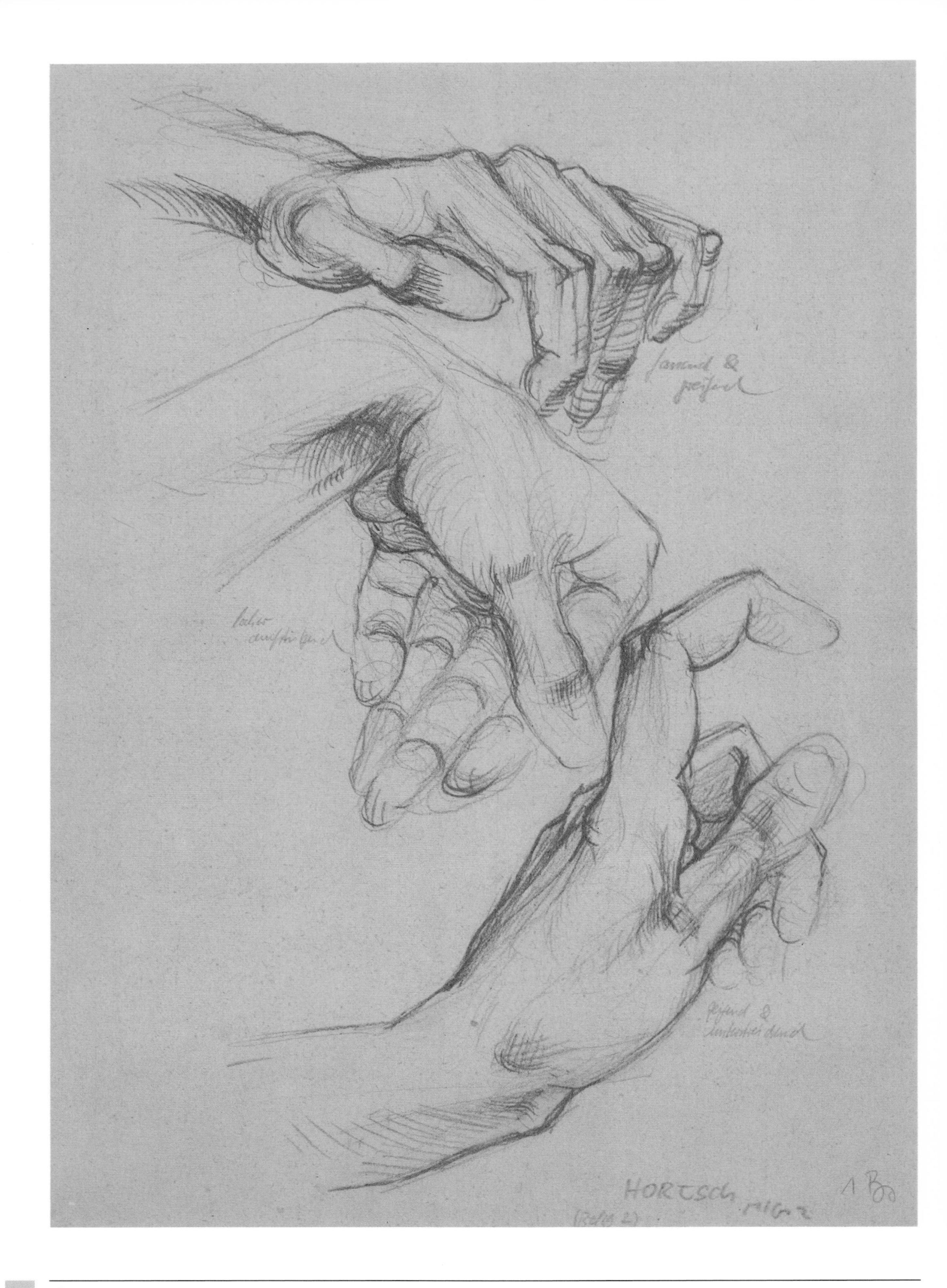

144 ARM FORM CONSTRUCTION

Ascertaining how the basic directions and forms of the muscle mass relate to each other must be the first consideration. The different directions in which the muscle masses extend and the way in which they lock together must also be worked out.
Student of painting/graphic art, fourth semester

145 THE UNITY OF ARM AND HAND GESTURES

Just as the arm can be distorted, turned or bent to an extreme degree, similar changes of shape occur in opening, spreading or laying down the hand.
Student of painting/graphic art, fourth semester

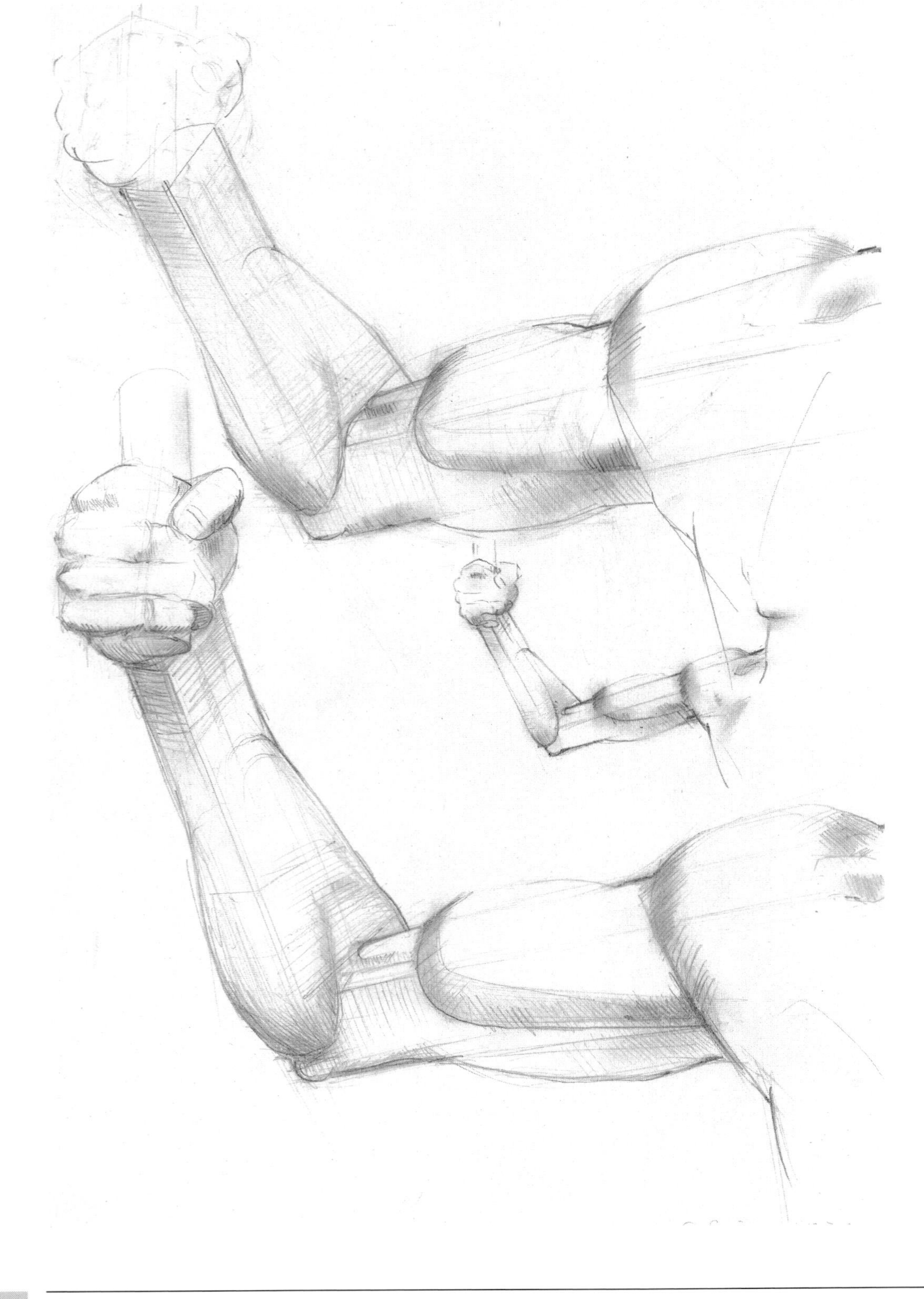

Hortsch, G.

146 HOW THE SPATIAL PLANES RELATE TO EACH OTHER
The pronounced faceting of the whole arm and hand down to the finger joints draws together the hatchings marking the spatial direction of the planes enclosing the body; the facets convey the many-sidedness of the arm's three-dimensional disposition.
Student of painting/graphic art, fourth semester

147 SIMPLE STUDIES OF HOW THE INDIVIDUAL PARTS FIT INTO A WHOLE
Indication of how the sectional changes in direction continue right through the joints also indicates the position of the whole arm and its elements in space.
Student of painting/graphic art, fourth semester

148 THE IMPACT OF FUNCTION ON THE MODELING OF THE ARM AND HAND
When the arm is leant on or supported on the hip, i.e. used for specific tasks, the muscles relate to the skeleton in a specific way and this is expressed in an a structural understanding of the form, supported by mental visualization of how cross-sections would run.
Student of sculpture, fourth semester

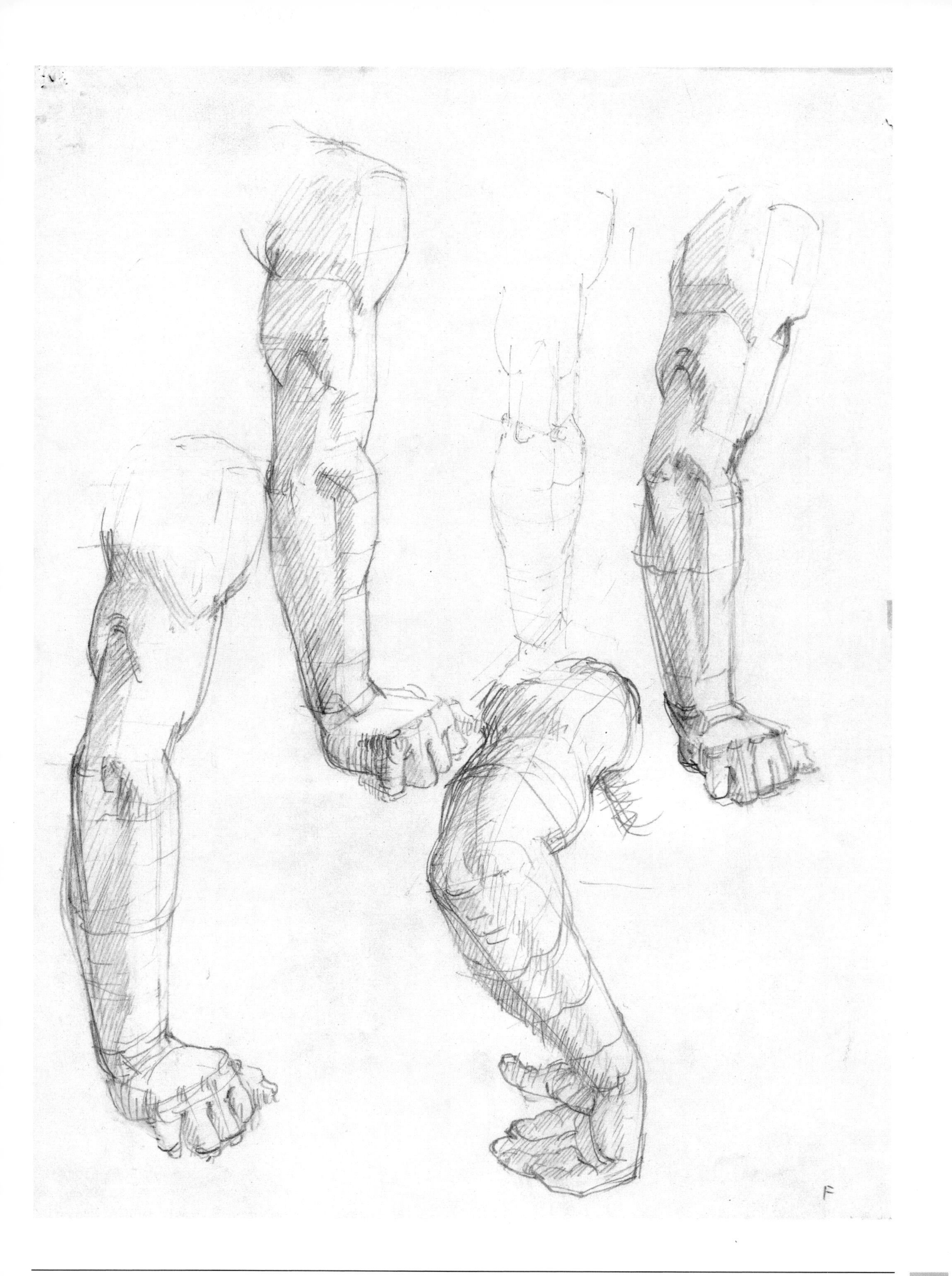

150 CONFIDENT UNDERSTANDING OF FORM – THE BASIS OF FREE EXECUTION

A free approach to investigating surface texture and structural diversity can succeed only on a foundation of acquired understanding.

Student of painting/graphic art, fourth semester

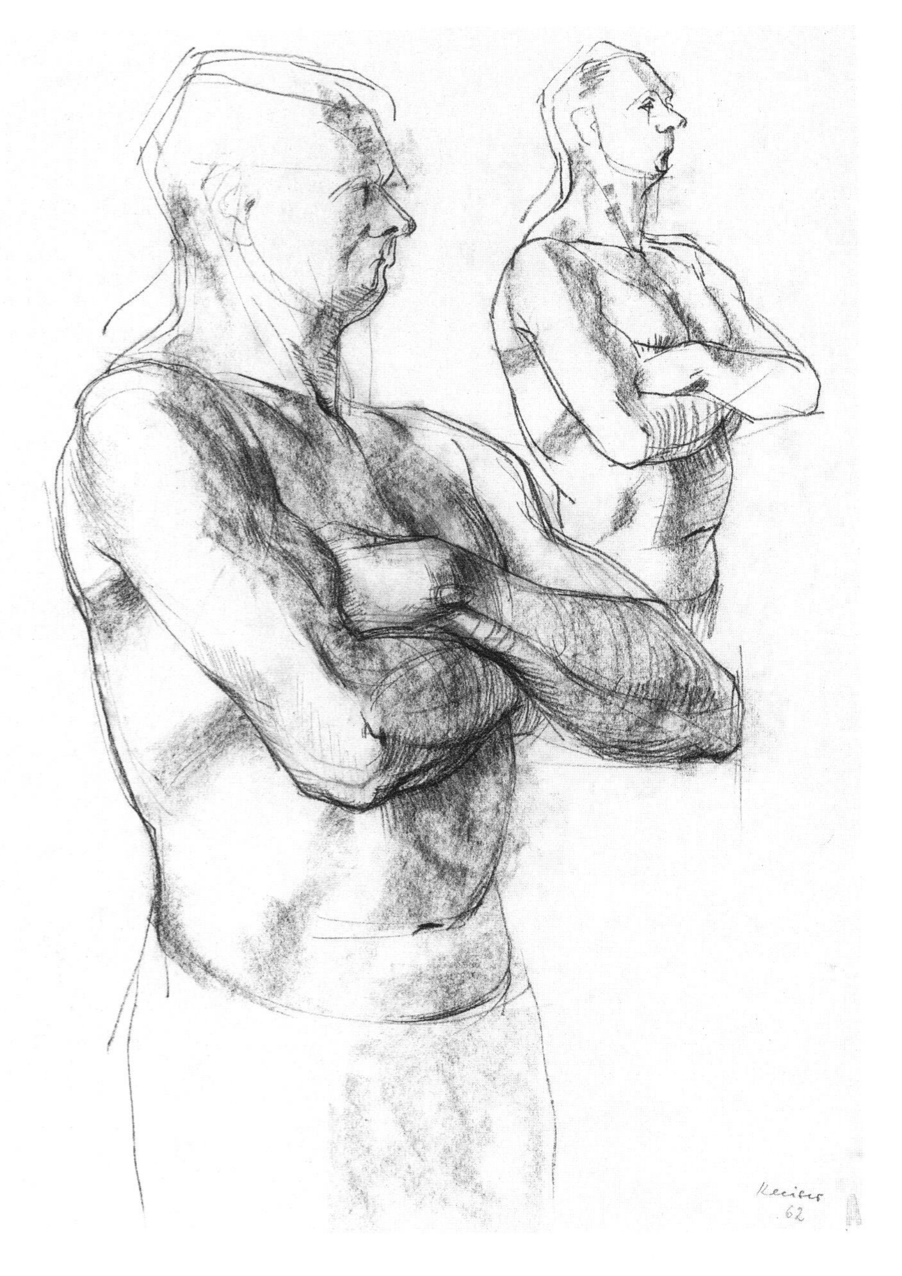

149 CONTRIBUTION MADE BY BASIC UNDERSTANDING OF FORM

The treatment of the arm mass reveals that the person drawing is alive to basic concepts such as the four-sidedness of the upper arm and the conical shape of the lower arm.

Student of painting/graphic art, fourth semester

PAGE 106

151 COMBINED STUDY OF THE ARM AND TORSO

The arm actions are ultimately derived from its connection with the torso. White highlights on a colored ground reinforce the plastic effects.

Student of restoration, fourth semester

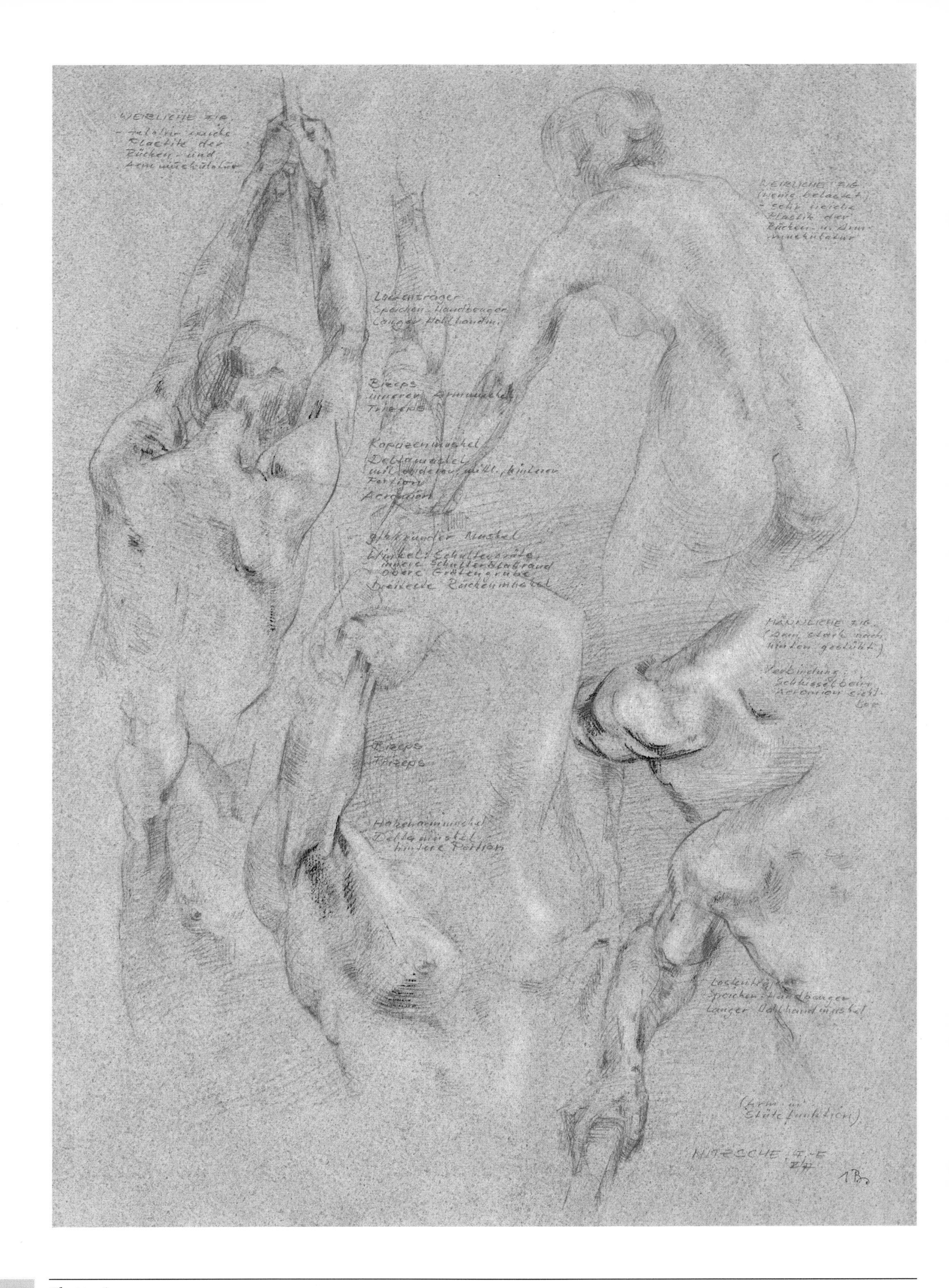

Chapter 8
Studies of the construction, functioning and plastic behavior of the hand and arm

9.

Studies of the whole figure

All the discussions and assignments described in previous chapters should be seen as necessary preliminaries for tackling the whole figure. It should be absolutely clear that it is not now simply a question of assembling the sections of the body so far treated separately. The objective of anatomical study for artists is to create indivisible and logical connections, until the point is reached where the student can invent freely (fig. 194). The studies now to be undertaken require graphic clarification of what is happening functionally and plastically, factors that crucially affect the whole figure in some of its basic situations. We are now concerned with:

- Standing poses in repose in their simple forms and in forms involving movement (figs 152–174), the problems of solidity in basic and three-dimensional views of the model from the strictly constructed study to the deliberately open, more impressionistic one.
- Seated poses in their basic forms and their manifold variations (figs 175, 176, 179–186) with the most important consequences of function, progressing again from graphic layout (figs 176, 179) by way of architectonic understanding (figs 180, 181) to free execution (figs 182–184).
- Tackling assignments involving two figures (figs 9, 15, 174, 177, 178, 185).
- Reclining poses with their great scope for change and the necessary functional characteristics (figs 187–191, 193, 195, 196).
- How to tackle free figural inventions based on visual imagination (fig. 194).

In a simple standing pose with the body weight distributed equally on both feet – in basic front and back views (figs 152, 156) – the difficulty lies in conveying the plasic cores in their forward-backward movement convincingly, because the curves of the spinal column cannot directly indicate the position of the pelvis, loins and rib cage in this position. The only recourse is to encompass the whole mass comprised by the pelvis and rib cage (fig. 156). Including a multiplicity of internal forms (fig. 152) cannot save the situation – the result is fragmentation. The choice of three-quarter views is therefore recommended, making it easier to locate characteristic solid and three-dimensional attributes (figs 153–159, 166).

Raising the arms vertically even in a static standing pose causes clear structural separation of the pelvis and rib cage (fig. 153). The behavior of the masses and the meeting of their forms can be deduced and organized by indicating where cross-sections would run (figs 156, 159, 166). Actions such as twisting the body in a standing pose (figs 169, 172) can be mastered only if you follow the spiral course of the central axis and show the upper body and pelvis in each case occupying different positions in space. Forming the habit of bearing cross-sections in mind is effective also when the structure of the body is kept open (figs 161, 172). In a contrapposto standing pose both the interplay of function and structure and supporting and supported forces must be made clear (figs 165, 167, 168, 174).

The layout of a seated pose also starts from the central axis of the body and the first concern is to make clear basic solid forms seen as constituent elements so that they can be used to control any foreshortening that may occur (fig. 176). In forward- or backward-leaning sitting positions (figs 175, 180), observation of the stretching or compression of the abdominal wall is required, and attention is needed to the behavior of the shoulder girdle when the model is leaning back and supported on his or her arms. A relaxed, free approach should always be aimed at while thinking through the construction of the seated figure (fig. 181). It is amazing how little information the eye needs to turn indications into a complete whole.

If two models are used side by side as objects for a study, you should ensure that at least several intersections occur to prevent the two figures falling apart from each other (important for sculptural assignments, figs 9, 16, 177, 178, 185).

Studies of intermediate positions between sitting and lying, in which different parts of the body (e.g. the arm, elbow and/or buttocks) take over supporting functions, have great charm. Here you must always consider the implications in the torso area when the loading on the two sides of the body is unequal (figs 186–188): what happens to the shoulder girdle when the arm or elbow is leant on, how the torso is stretched on the supported side, what effect the telescoping on the unsupported side has on the soft, fleshy forms.

Apart from the caving in of the abdominal wall, a simple lying pose with the weight supported equally by the back and buttocks affords very little functional expression (fig. 190). On the other hand surprising form outlines occur when the pelvis and rib cage are twisted (side position of the pelvis, prone position of the rib cage, figs 189, 191, 193). There are steep inclines in the pelvis, and flat shapes in the area of the upper body, while the movement implications can be seen in the middle area of the torso. The distortions in the abdominal wall can be conveyed expressively only if you are absolutely clear as to the spatial position of the plastic cores visualized as individual elements. Then the behavior of the soft, fleshy forms of the stomach or buttocks becomes an inevitable logical consequence. In the female nude model this also applies to the behavior of the breasts: when the model lies on her back they sink towards one another under their own weight; if the arms are clasped behind the head the breasts must follow the arms (fig. 193); and if the body is turned towards the stomach and the upper body supported they will hang down (fig. 191).

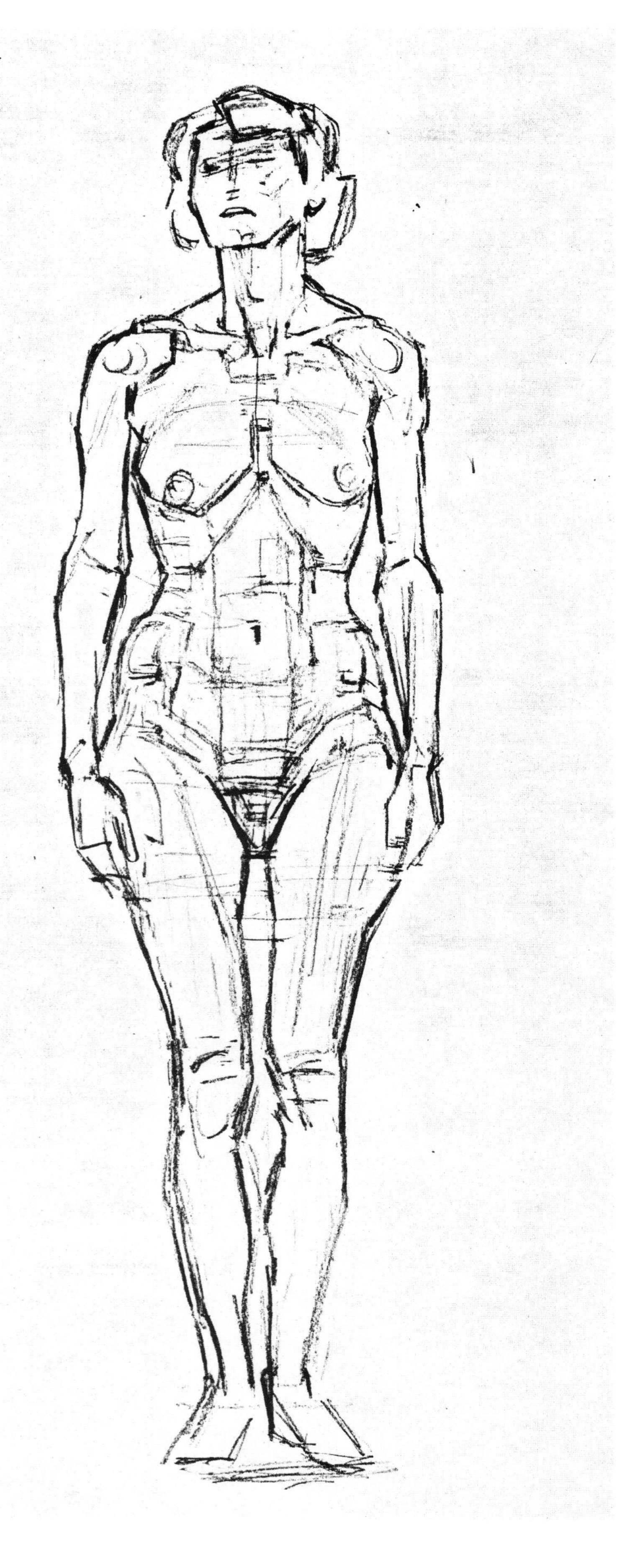

152 THE PROBLEMS THE VIEWING ANGLE CAN CAUSE IN CONVEYING SOLIDITY AND THREE-DIMENSIONALITY

In a basic view of the whole figure it is difficult for the learner to recognize the rhythmic course of the central axis of the body and consequently of the various body masses. In this view it is hard to show attributes of form, e.g. the way the upper and lower body can move in relation to one another, and even introducing a large number of internal forms is of no great assistance.

From a Bammes course at the Schule für Gestaltung, Zurich

153 MAKING DRAWING EASIER BY MEANS OF THE MODEL'S ACTIONS AND A MORE FAVORABLE VIEWING ANGLE

The search for expression in the behavior of the body in a standing pose is helped by the choice of a slightly offset frontal view and the emphasis of the breasts and pelvic mass resulting from the raising of the upper body. The way the masses are clearly contained and grouped can be credited to the artist's skill.

Student of painting/graphic art, fourth semester

1- Bo
Mosler · BÜ II

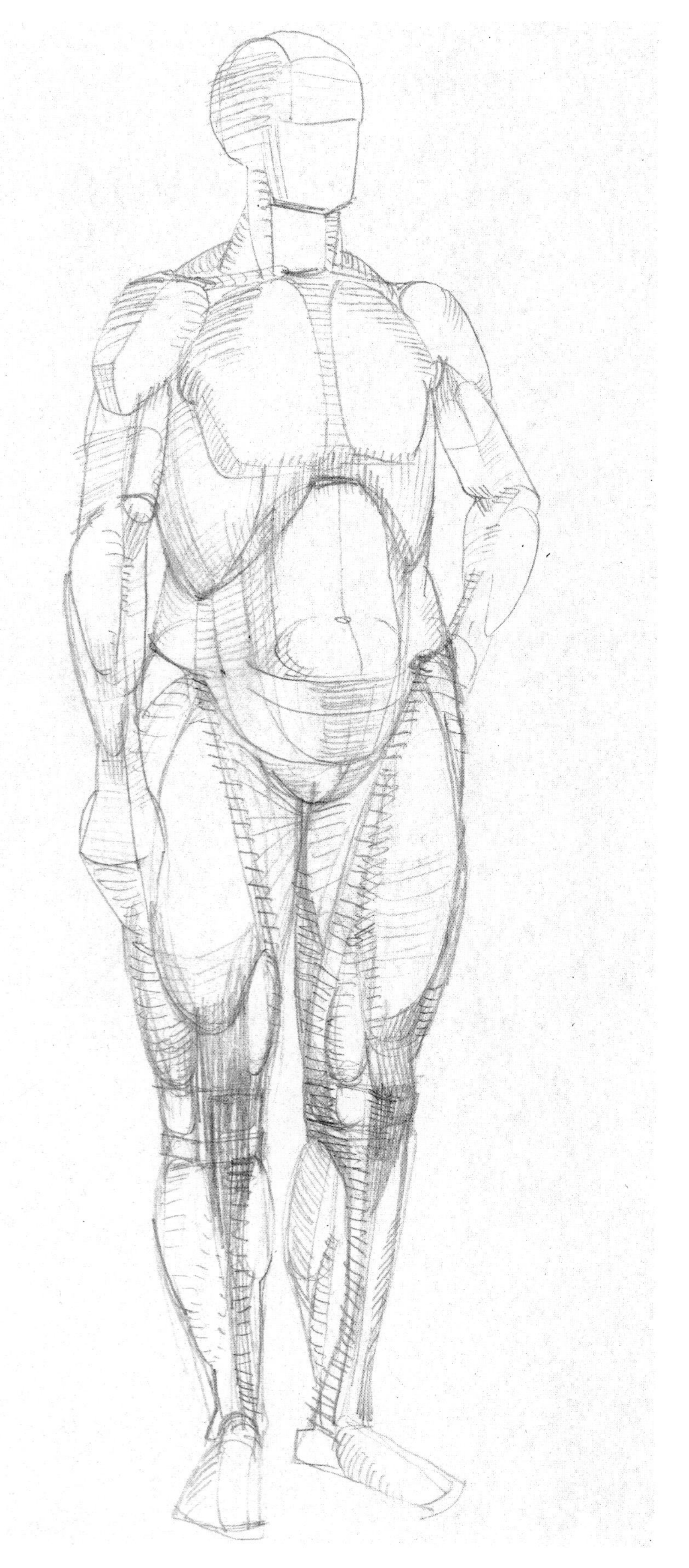

154 STRUCTURALLY ORGANIZED DRAWING IN A NEAR FULL-FRONTAL VIEW

The artist here is using his concepts of the details of the structural blocks, especially around the rib cage, pelvis and knees. No opportunity is neglected to fill in all aspects of the figure visible from the front, and to a small extent from the side. Forms are coarsened to powerful effect.
Student of set painting, fourth semester

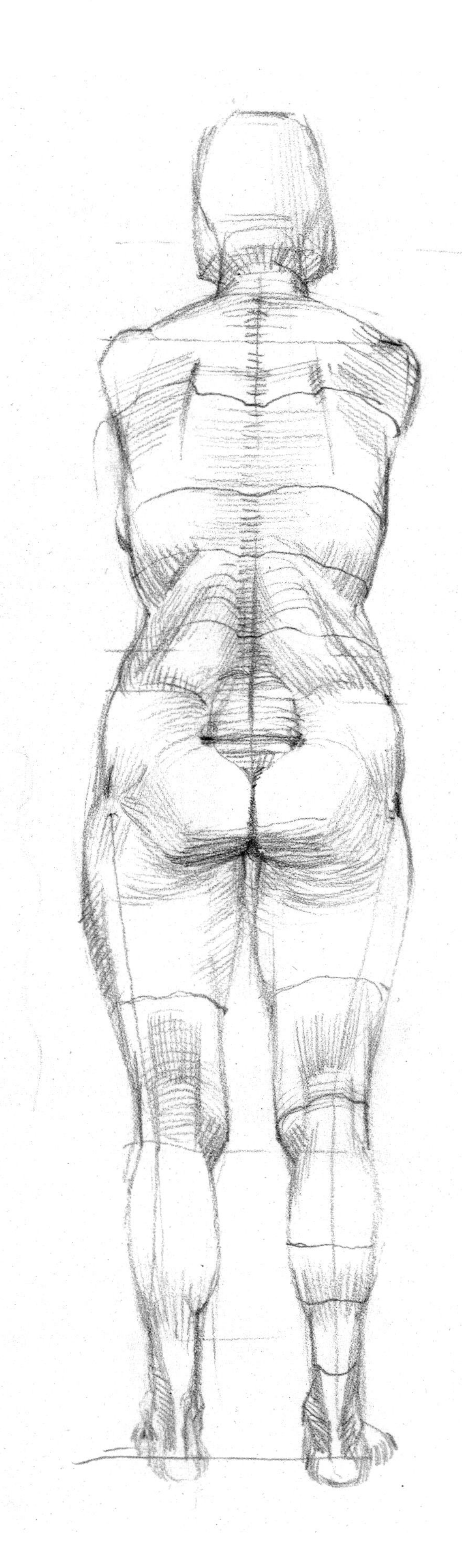

155 ATTEMPT TO RESOLVE THE PROBLEMS OF A NEAR-BACK VIEW

The most important decision made here is the endeavor to establish the angles of the planes of the body following the course of the spinal column.

Student of painting/graphic art, fourth semester

156 CONVEYING A FULL BACK VIEW IN A STANDING POSE

The difficult graphic task of conveying the relatively inarticulated back view is tackled by concentrating on the way the spinal column recedes near the loins and the neck, and on the contrast between the convexity of the rib cage and the intersections through the area of the loins.

Student of painting/graphic art, fourth semester

157 FORMAL FEATURES ARE EASIER TO GRASP IN ANGLED VIEWS

Diagonal views display a wider range of formal features and make them less ambiguous, which is an advantage to the person drawing.

From a Bammes course at the Schule für Gestaltung, Zurich

158 SUCCESSFULLY CONVEYING THE ESSENTIAL TRAITS OF THE FIGURE

This half back view is the work of the same student who drew fig. 152; with the unequivocal definition achieved by the viewing angle, crucial features of the interrelationships between the main shapes of the body are discovered, the movement in the form of the spinal column is recognized and the psychological aspect of the overall pose is also captured.

From a Bammes course at the Schule für Gestaltung, Zurich

159 GIVING WEIGHT AND SOLIDITY TO THE POSE

Once unequivocal viewing angles (figs 157, 158) and the expression of the overall pose have been mastered, further differentiations of individual volumes and how they meet become possible.

Student of sculpture, fourth semester

160 PART OF AN EXAMINATION ASSIGNMENT SPECIFICALLY SET FOR STUDENTS OF RESTORATION

It is important that restorers should be trained to fill in 'missing bits' of works of art by copying the drawing of a master (here Schnorr von Cardsfeld) and using their anatomical skills to prepare muscle and skeleton analyses of it.

Student of restoration, end of fourth semester

161 EXAMINATION ASSIGNMENT SPECIFICALLY SET FOR STUDENTS OF PAINTING/GRAPHIC ART

The value of the assignment lies in demonstrating that the student's skills and knowledge of anatomy can be used in a free, open, economical approach.

Student of painting/graphic art, end of fourth semester

162 WORK PARTLY BASED ON VISUALIZATION IN AN EXAMINATION ASSIGNMENT

The model's pose was held only briefly, and had to be assimilated as a basis for analyzing how the skeleton form and the main muscle groups combine to make up the architecture of the body. All that had been learned previously was tested here.

Student of stage design, end of fourth semester

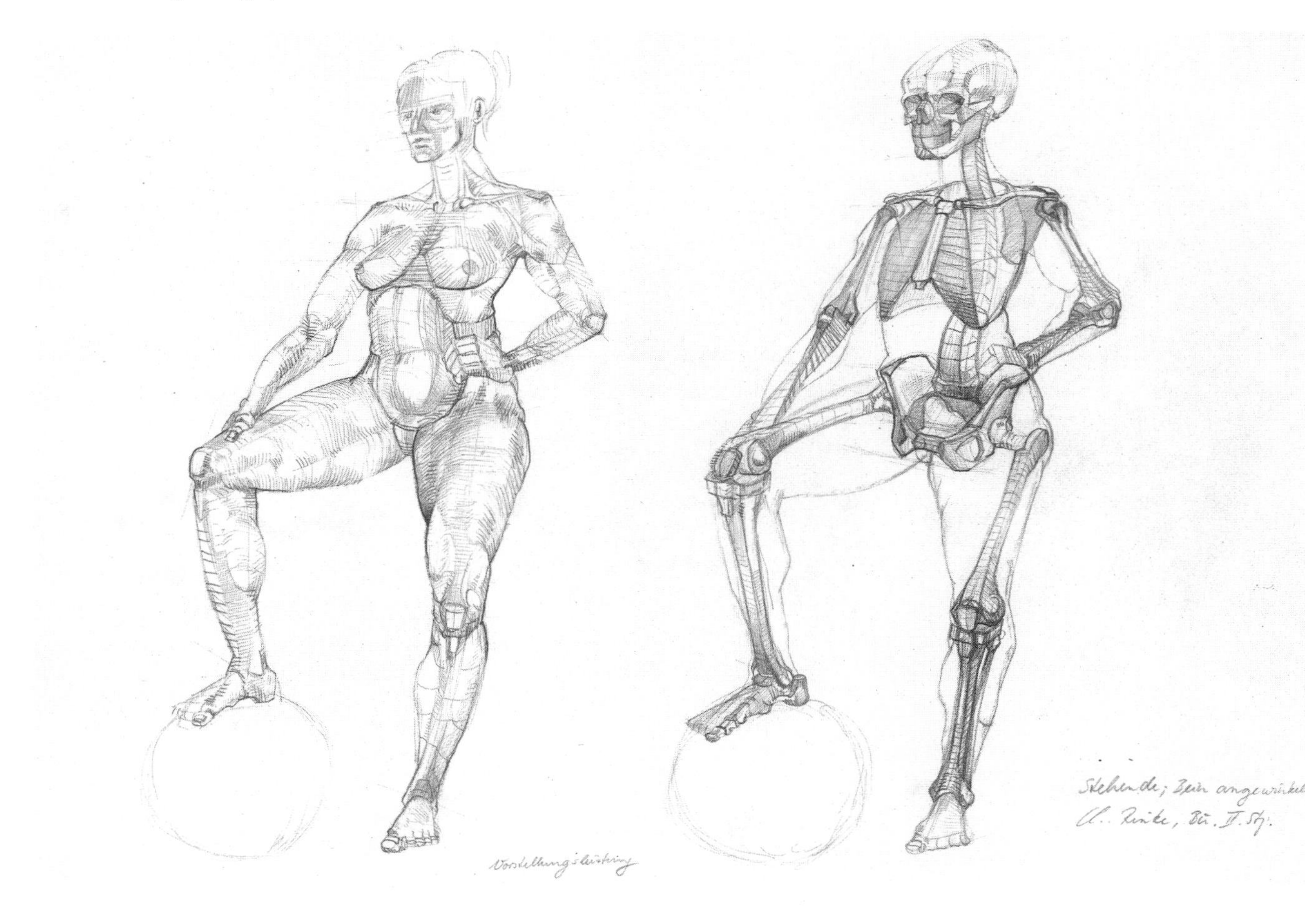

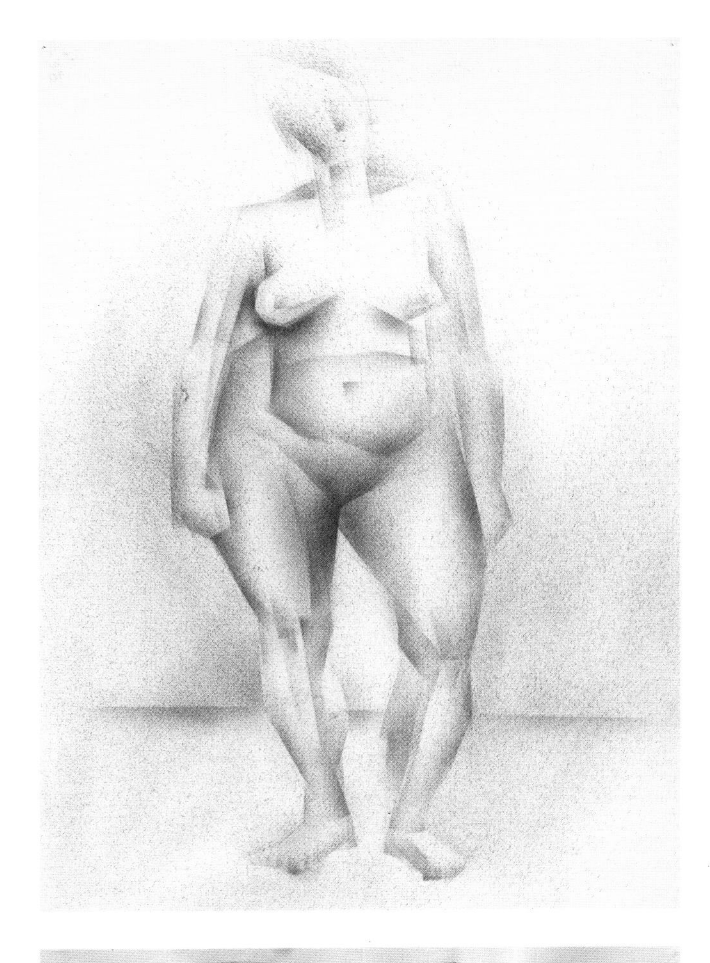

163 A VARIANT APPROACH TO EXPRESSING MASSIVENESS OF VOLUME

Using an airbrush an attempt has been made to express the large curve of the load-bearing hip and large, opulent body masses in various sequences where the forms were masked or revealed.

Student of set painting, fourth semester

164 TRYING TO FOLLOW THE FLOW OF MOVEMENT, USING A THREE-DIMENSIONAL AND A TWO-DIMENSIONAL APPROACH

To follow through a movement of the whole form, it is sometimes enough to capture the course of the movement using a medium which does not tempt the student to pile on detail.

From the Salzburg intensive course, 1988

165 HOW FUNCTIONAL STRESSES WORK IN A CONTRAPPOSTO POSE

The interchange between bearing the weight and not doing so, stretching and compression, is the main subject of this study, with the planes that are to view being carefully introduced.

Student of painting/graphic art, fourth semester

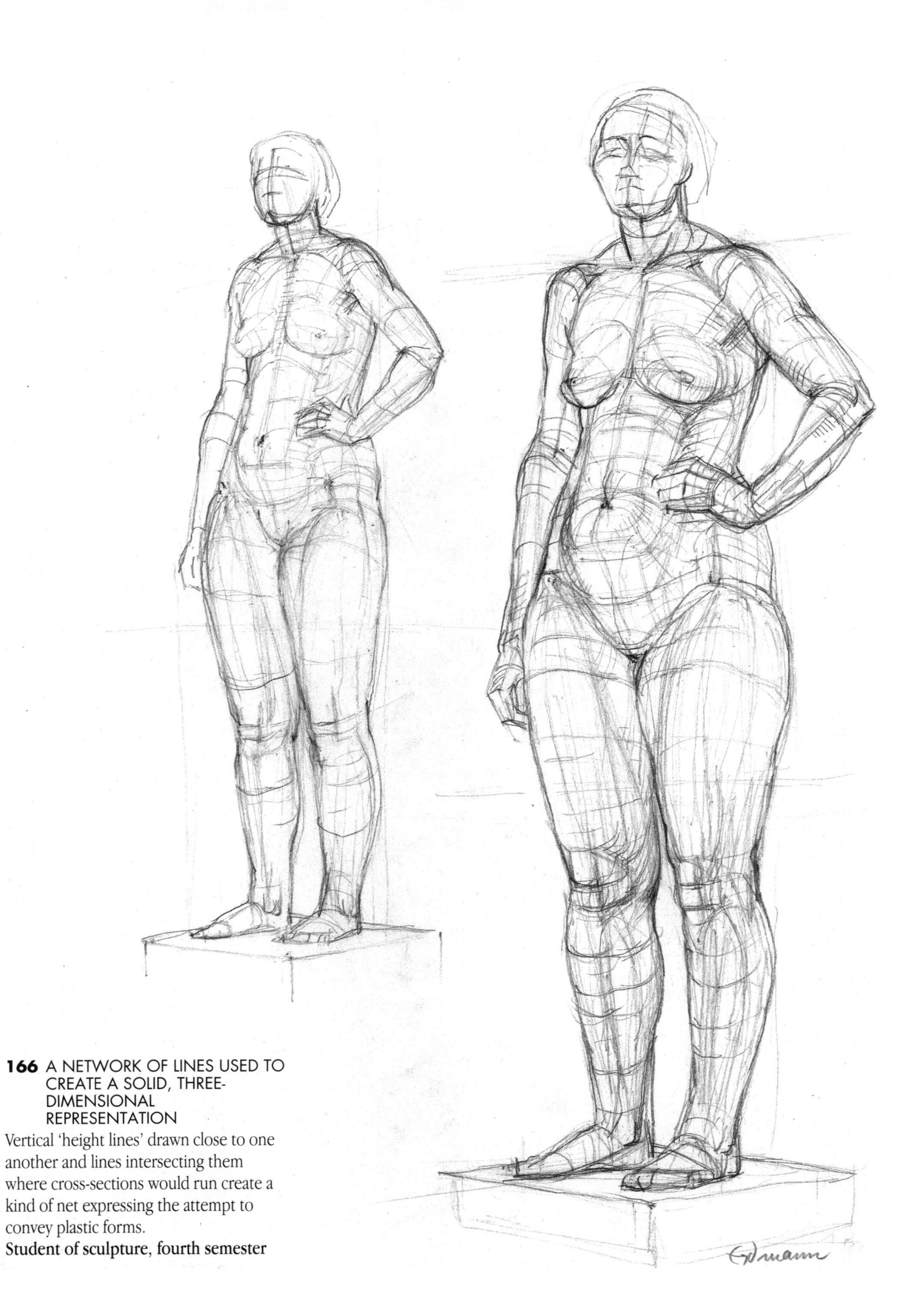

166 A NETWORK OF LINES USED TO CREATE A SOLID, THREE-DIMENSIONAL REPRESENTATION

Vertical 'height lines' drawn close to one another and lines intersecting them where cross-sections would run create a kind of net expressing the attempt to convey plastic forms.

Student of sculpture, fourth semester

zwicker

167 PAINTERLY USE OF METHODS EMPHASIZING SURFACE TEXTURE

In total contrast to the structural drawing in fig. 166, the student here elaborates the impression of the softness and curves of the surface of the body using a range of dynamic tonal gradations.

Student of painting/graphic art, fourth semester

168 EXAMINING PLASTIC EFFECTS IN A CONTRAPPOSTO POSE

In contrast to figs 166 and 167, the student here concentrates on indicating the main surface planes and developing the body in depth by using hatching of varying density, especially in the loin area where the spatial depth is contrasted with the curve of the back and buttocks; the compressed masses on the side of the weight-bearing leg are brought closer together.

Student of painting/graphic art, fourth semester

169 CONTRASTING VIEWS OF THE BODY WHEN THE TORSO IS TWISTED IN A STANDING POSITION

The frontal view of the rib cage combined with the three-quarter view of the pelvis causes a spiral twist in the line of the stomach. Finally the arms are drawn at a 90 degree angle to the position of the legs.

Student of sculpture, fourth semester

170 WORK PARTLY BASED ON VISUALIZATION ANALYZING EXPRESSION OF MOVEMENT

Both figures are components of a three-part examination assignment; students were asked to prepare analyses of the skeleton, muscles and live appearance, in this case of a ball-throwing pose which the model held only briefly. The objective was to express movement with total conviction.

Student of painting/graphic art, end of fourth semester

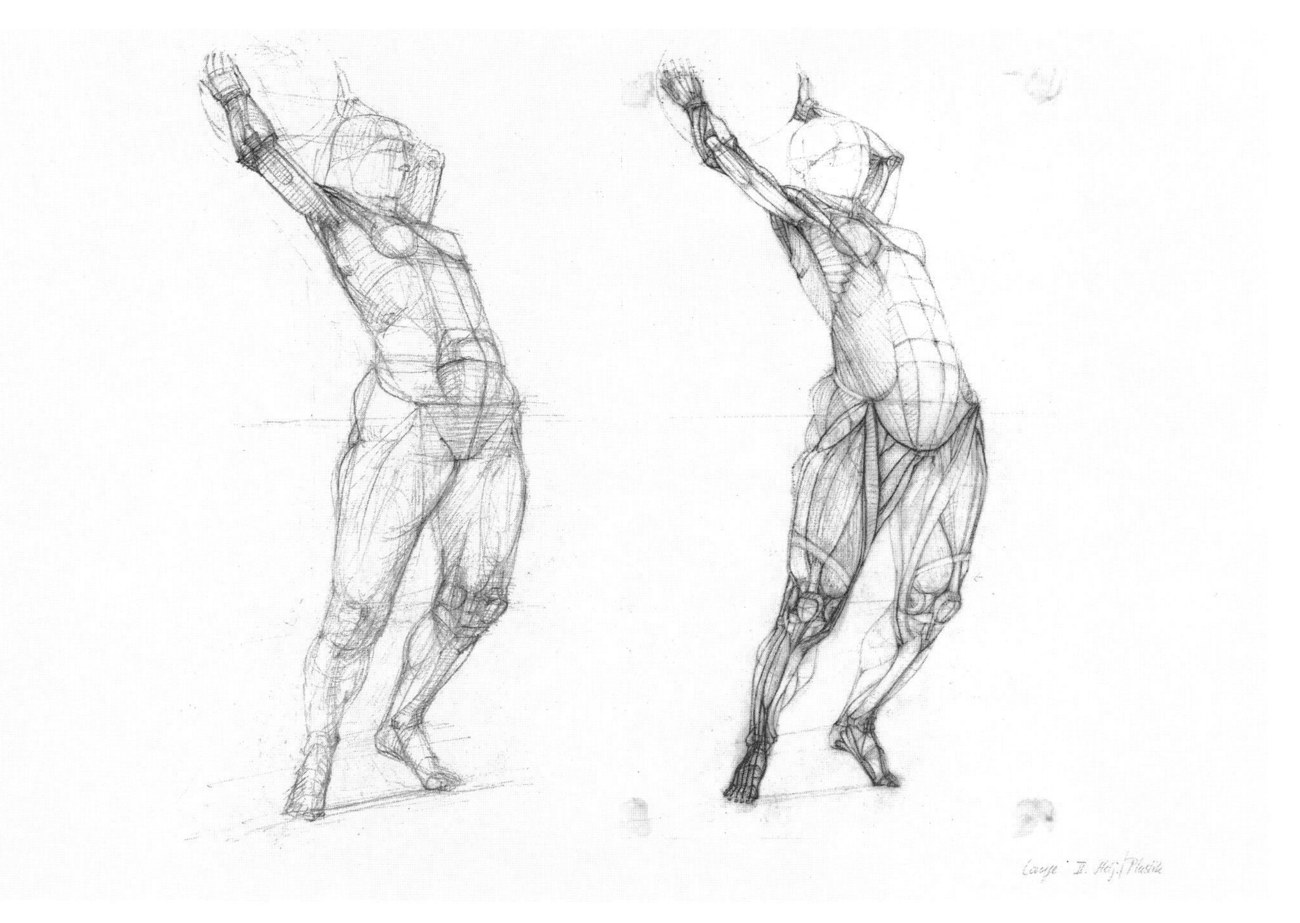

171 CATCHING THE EXPRESSION OF MOVEMENT USING A KETCHING TECHNIQUE

However quickly a study of movement may have been executed, in this case conveying the essential processes involved in twisting and bending the torso and supporting processes along with spatial factors, it must be accurate.
Student of sculpture, fourth semester

172 LIMITED OBJECTIVES AND SIMPLIFYING THE APPROACH

This study concentrates mainly on the directions of the visible body planes using corresponding areas of hatching and blank areas. Limiting one's objectives and simplifying the approach are not mutually exclusive.
Student of painting/graphic art, fourth semester

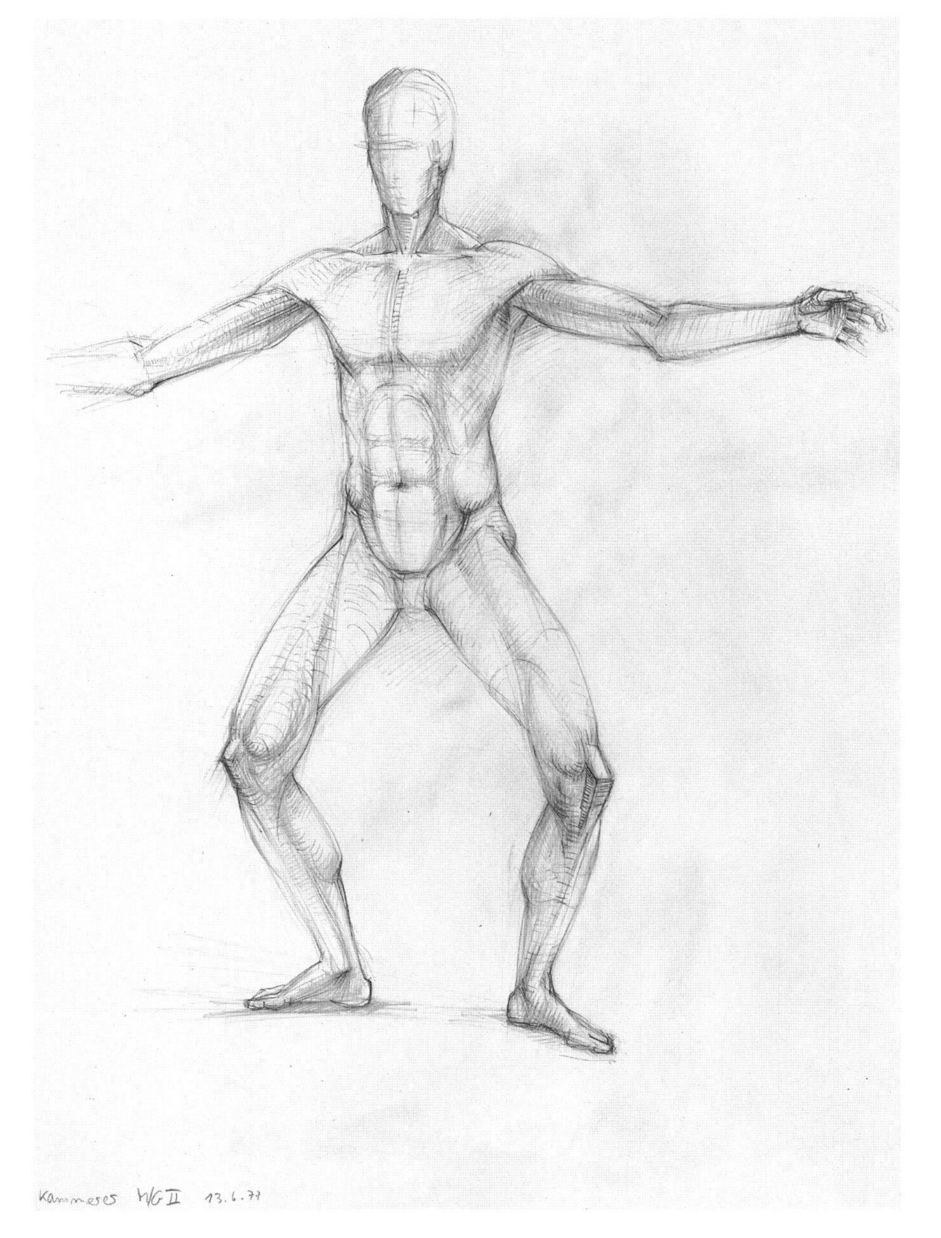

173 THOROUGH CONSIDERATION OF FUNCTIONAL PROCESSES

Before the graphic study of the dancer's pose (held only briefly) could begin, students had to give a verbal description of all the functional and modeling processes that might be expected; only then were they able to see their functional predictions confirmed by the model.

Student of painting/graphic art, fourth semester

174 A STUDY ASSIGNMENT WITH TWO FIGURES BASED ON VISUALIZATION

The drawing of these figures is based on the principles involved in understanding the architecture and plastic behavior of the body. Students were free to choose their own means of presentation.

Student of sculpture, fourth semester

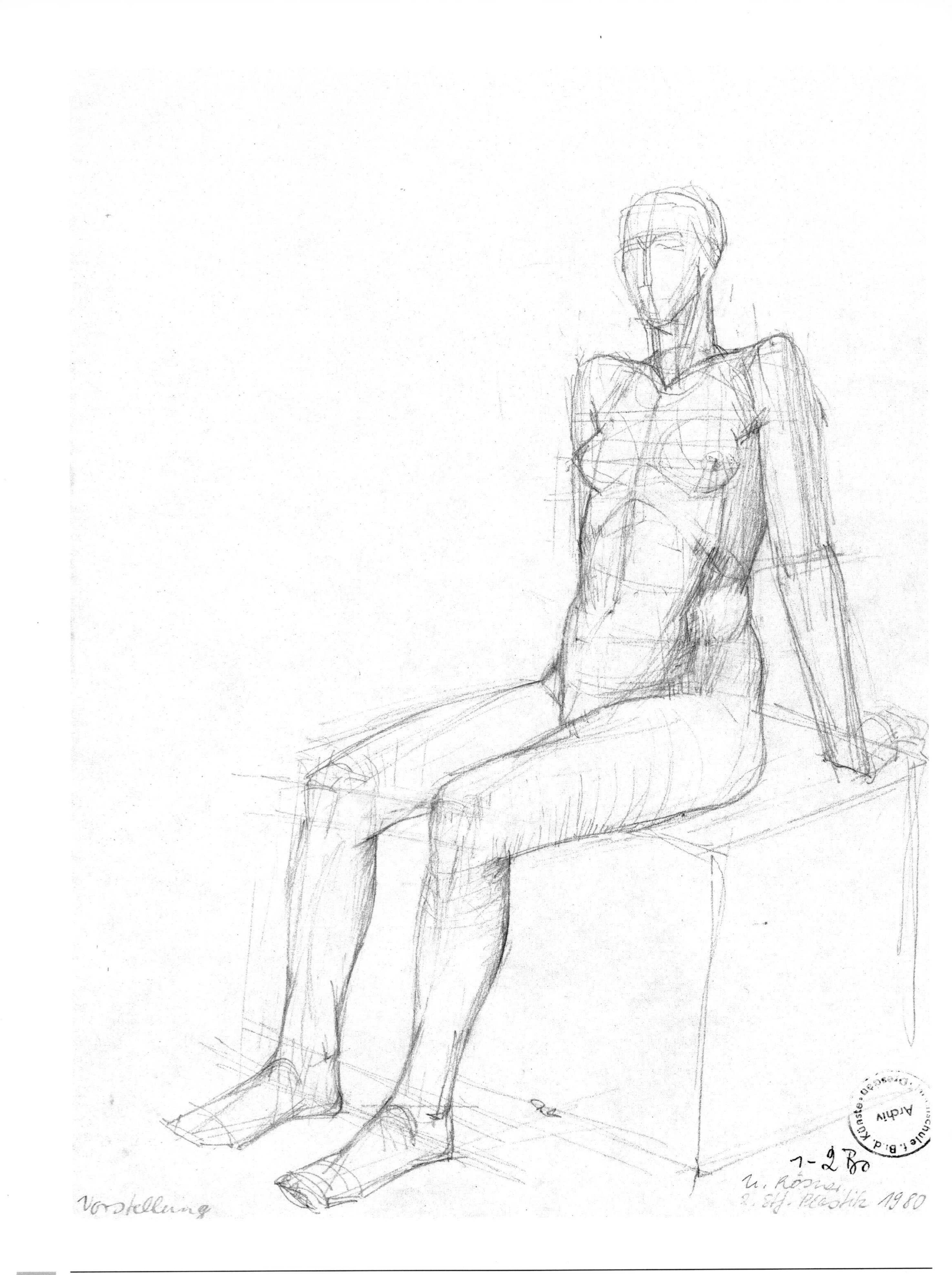
Vorstellung
Archiv

175 APPLYING THE THREE-DIMENSIONAL REFERENCE SYSTEM TO A SIMPLE SEATED POSE

Both functional and three-dimensional factors are established by the skillfully inserted line indicating the course of the central axis of the body. The spatial relationships between the two sides of the body separated by the central axis are established by the cross-axial lines.
Student of sculpture, fourth semester

176 LAYOUT FOR A SEATED POSE INVOLVING MOVEMENT USING STRUCTURAL SHAPES SHOWN AS INDIVIDUAL ELEMENTS

To give valid expression to the way the upper body recedes from and the legs come toward the viewer, it is often essential to project elementary shapes (cube, sphere, half-sphere, cylinder) in their correct position and spatial location.
Student of sculpture, fourth semester

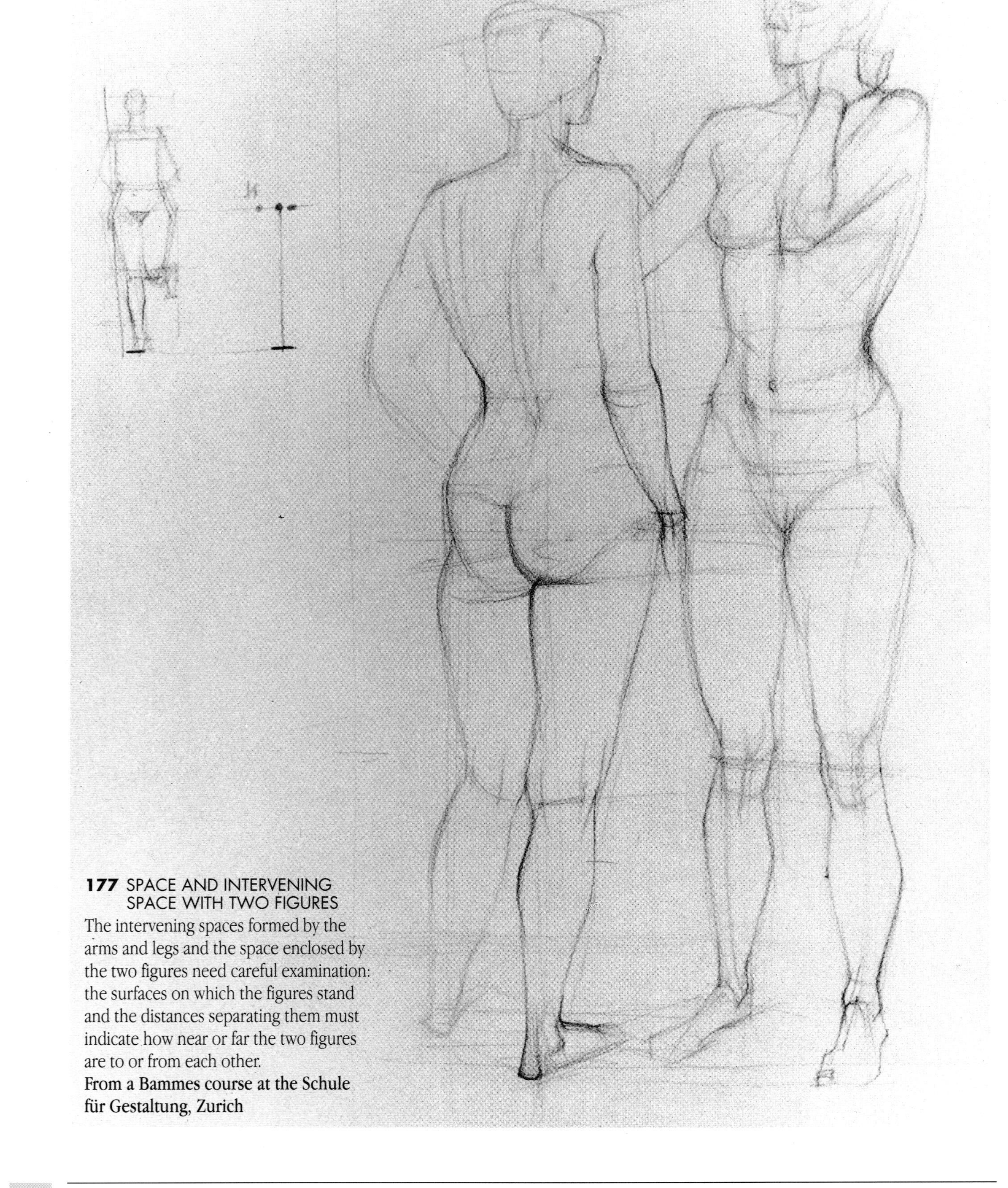

177 SPACE AND INTERVENING SPACE WITH TWO FIGURES
The intervening spaces formed by the arms and legs and the space enclosed by the two figures need careful examination: the surfaces on which the figures stand and the distances separating them must indicate how near or far the two figures are to or from each other.
From a Bammes course at the Schule für Gestaltung, Zurich

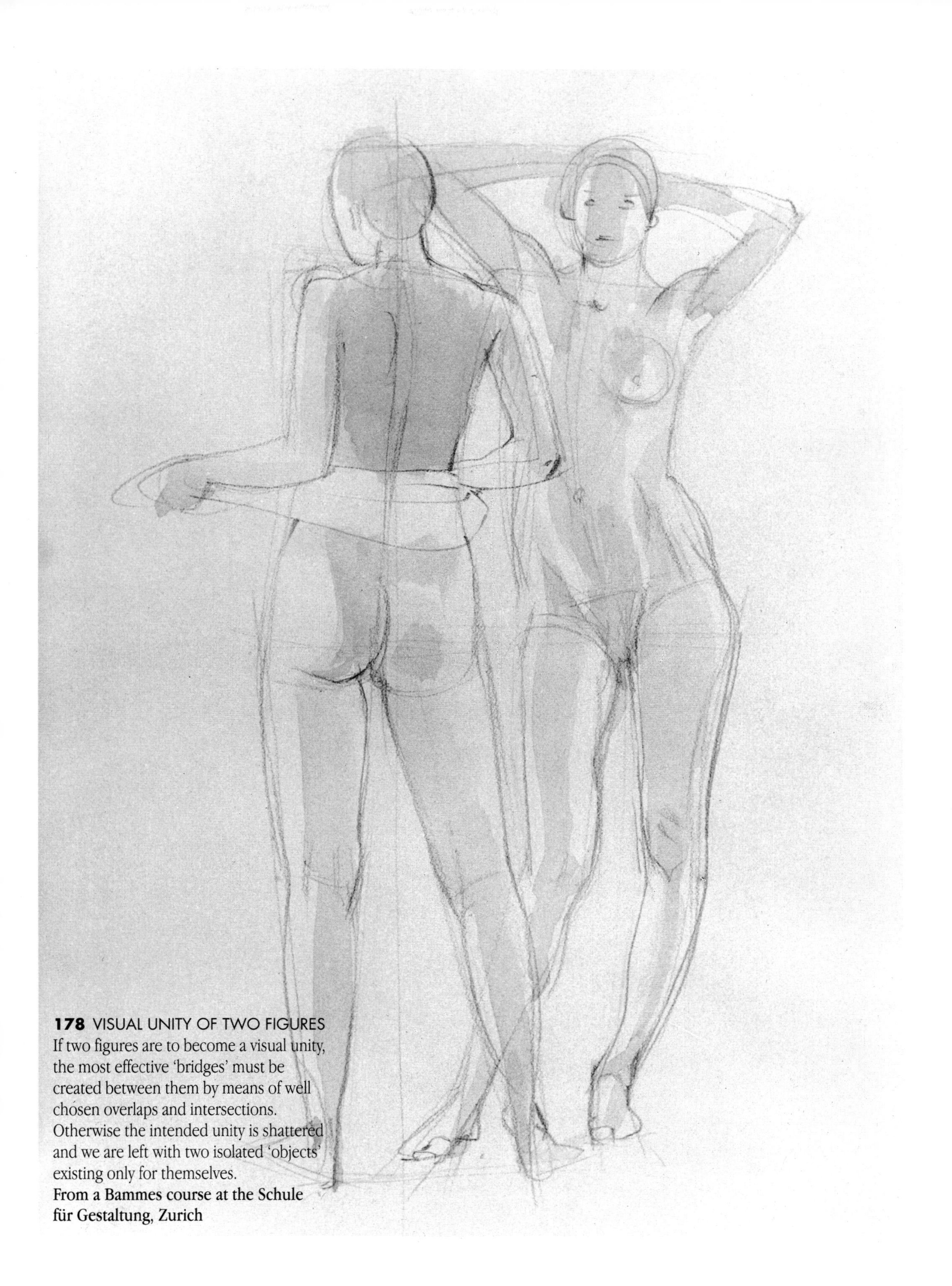

178 VISUAL UNITY OF TWO FIGURES
If two figures are to become a visual unity, the most effective 'bridges' must be created between them by means of well chosen overlaps and intersections. Otherwise the intended unity is shattered and we are left with two isolated 'objects' existing only for themselves.
From a Bammes course at the Schule für Gestaltung, Zurich

179 LAYOUT FOR A COMPLEX SITTING POSE WITH MULTIPLE FUNCTIONS

The position where the body is leaning back supported on one arm with the pelvis lying on its side requires first and foremost a definite idea of the directional lines and the angles formed by them, and secondly visualization of the basic body components. Only then can further development be undertaken.

180 DEVELOPING VOLUME, SOLIDITY AND FUNCTION

The primary decisions regarding graphic lay-out demonstrated in fig. 179 also preceded this study, enabling the internal spaces within the figure and the relationship of the limbs to each other and to the torso to be ascertained, so enlivening the construction of volumes, solidity and functions (leaning forward and support).

Student of sculpture, fourth semester

1-2 Bo
Erfreulich :
U. Rösner
2. Stj. Plastik 1980
Vorstellung

181 SEATED POSE INVOLVING MOVEMENT BUILT UP CONSTRUCTIONALLY

All the criteria covered in requisite preliminary drawings of the type shown in figs 179 and 180 create a dependable framework into which detail can be fitted without the study degenerating into an accumulation of details.

Student of set painting, fourth semester

182 COMBINATION OF VISUAL MEMORY AND CONCEPTUAL ABILITY

The model's seated poses and forms had previously been exhaustively discussed, with no drawing allowed; then in the absence of the model students were free to draw using recent optical experience, visual memory and their conceptual ability. The graphic use of outline and open areas produces an attractive and expressive effect.

Student of painting/graphic art, fourth semester

Gedächtnisleistung

183 CONNECTIONS BETWEEN THE BODY AND SPACE

Extending the study to linking the body and the surrounding space – although it is outside the field of anatomy for artists – demonstrates the potential for cross-fertilization with artistic life drawing. The figure recedes from the foreground into the farther layers of space, with the network of lines becoming more and more dense. The occasional contour lines were not inserted until the very end.
Student of painting/graphic art, fourth semester

184 TEST INVOLVING LARGELY FREE INVENTIONS

The method of graphic presentation was left to the students' discretion, and the exercise was based on the same pose as fig. 183; but as well as depicting the given view, students were required to produce drawings of differing views from imagination.
Student of painting/graphic art, fourth semester

185 DEPICTING THE BODY WITH ONLY THREE MAIN TONAL VALUES

To prevent the exercise with two figures from degenerating into an accumulation of detail and the intended picture space from being broken up, only three tonal values could be used: light (= paper ground), medium and dark.

Amateur artist from the Special School of Painting and Graphic Art

186 PLASTIC AND FUNCTIONAL STUDY

The transitional pose between sitting and lying requires observation of many factors: the supporting power of the arm, the way the body is suspended beside the supporting arm (how the shoulder girdle behaves), the tipped-up lying position of the pelvis and the compression folds on the stomach wall.

Student of set painting, fourth semester

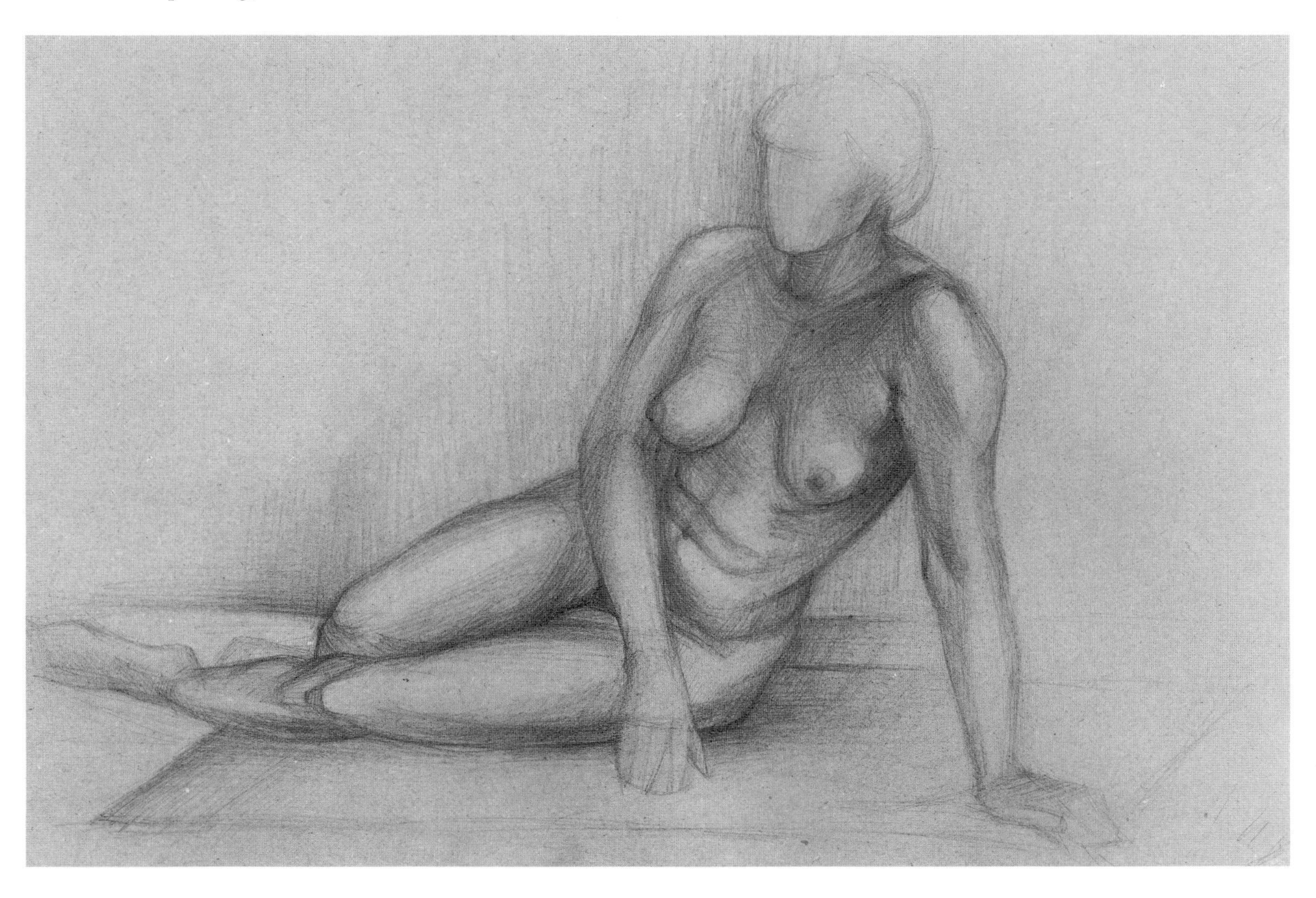

187 ANALYTICAL PENETRATION OF A RECLINING POSE

Although anatomical data are always clarified before work from life starts, now and then – as here in an examination assignment – difficult poses need to be clarified by analysing the muscles.

Student of painting/graphic art, fourth semester

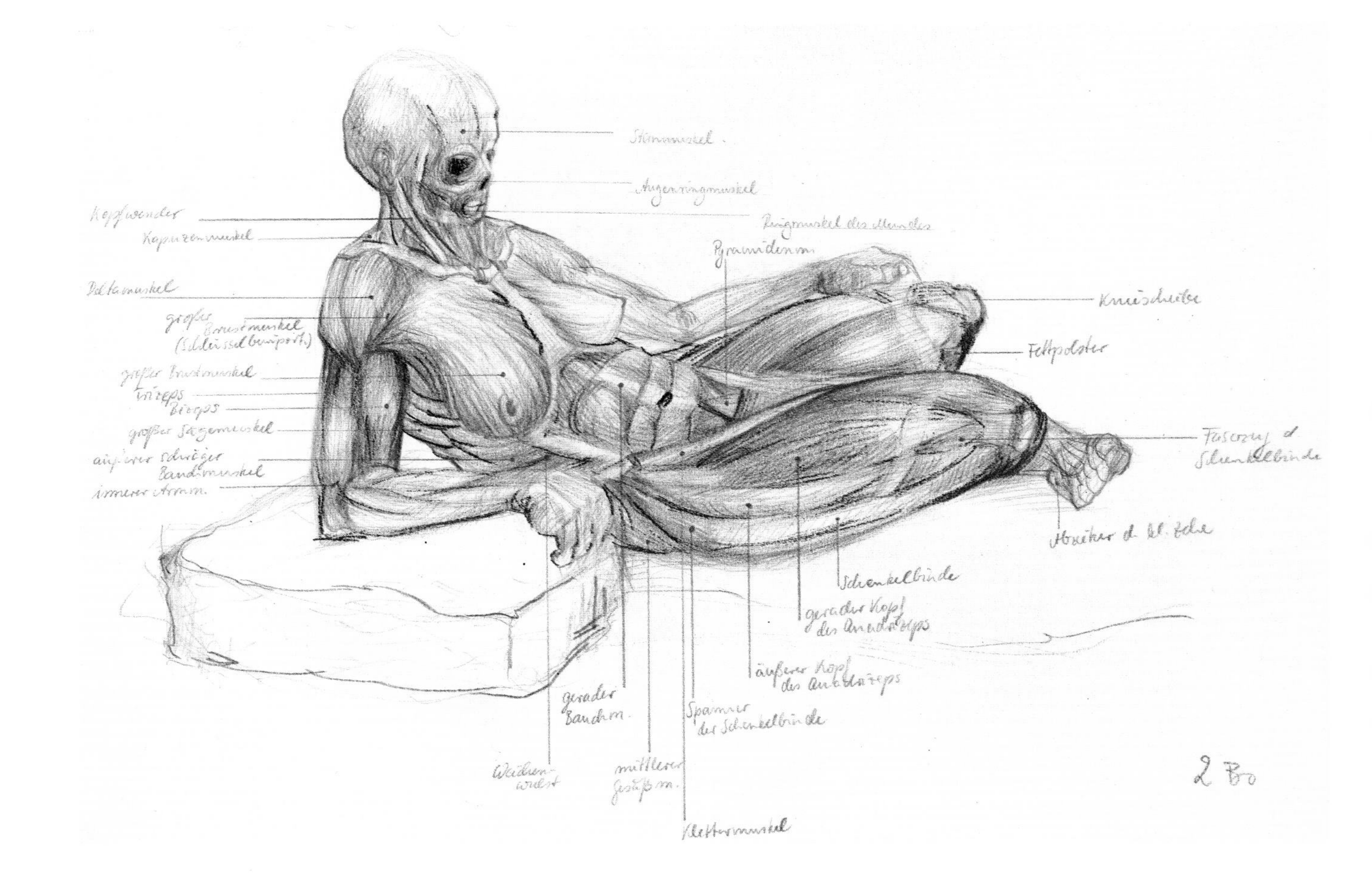

188 CONTRIBUTION OF THE LIVING IMPRESSION

For all our academic evaluation of anatomical factors, the study must not lose the living impulse of something seen and experienced, otherwise it withers into a dry technical drawing.

Student of painting/graphic art, fourth semester

189 ACHIEVING PLASTICITY BY INVESTIGATING CROSS-SECTIONS AND USING A WASH TO STRENGTHEN THE DRAWING

Investigating the changing dimensions of body mass is particularly relevant to sections of the body seen in extreme foreshortening. Washes applied with a brush can both bring out the facets of breaks between planes, highlighting the modeling, and to some extent merge the intrusive cross-section indications.
Student of painting/graphic art, fourth semester

190 PECULIARITIES OF FORM IN A SIMPLE RECLINING POSE

When the weight is carried equally by the pelvis and rib cage in a prone position, the student must pay attention to the way the abdominal wall caves in between the two bony centers and the breasts are flattened by their own weight.

Student of painting/graphic art, fourth semester

191 STUDY OF HOW FORMS BEHAVE IN A RESTING POSITION INVOLVING MOVEMENT

The pelvis jutting high above the model's flank and the supported rib cage turned toward the floor are the cause of unusual functional-plastic and sculptural changes, for example the 'screwing up' of the abdominal wall, the sagging of the soft under-belly and resulting projection of the crest of the hip bone, and the hanging shape of the breasts.

Student of sculpture, fourth semester

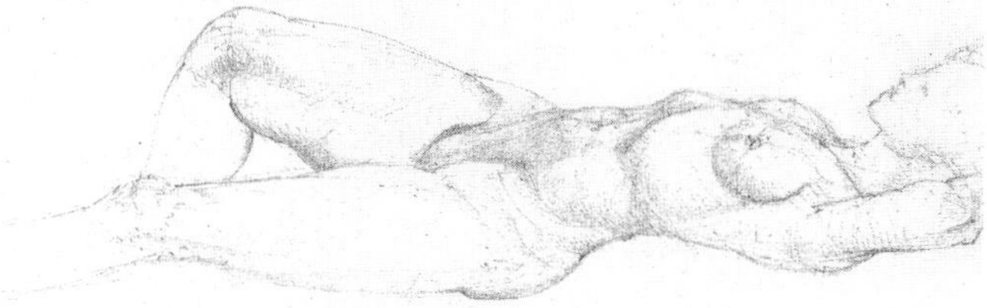

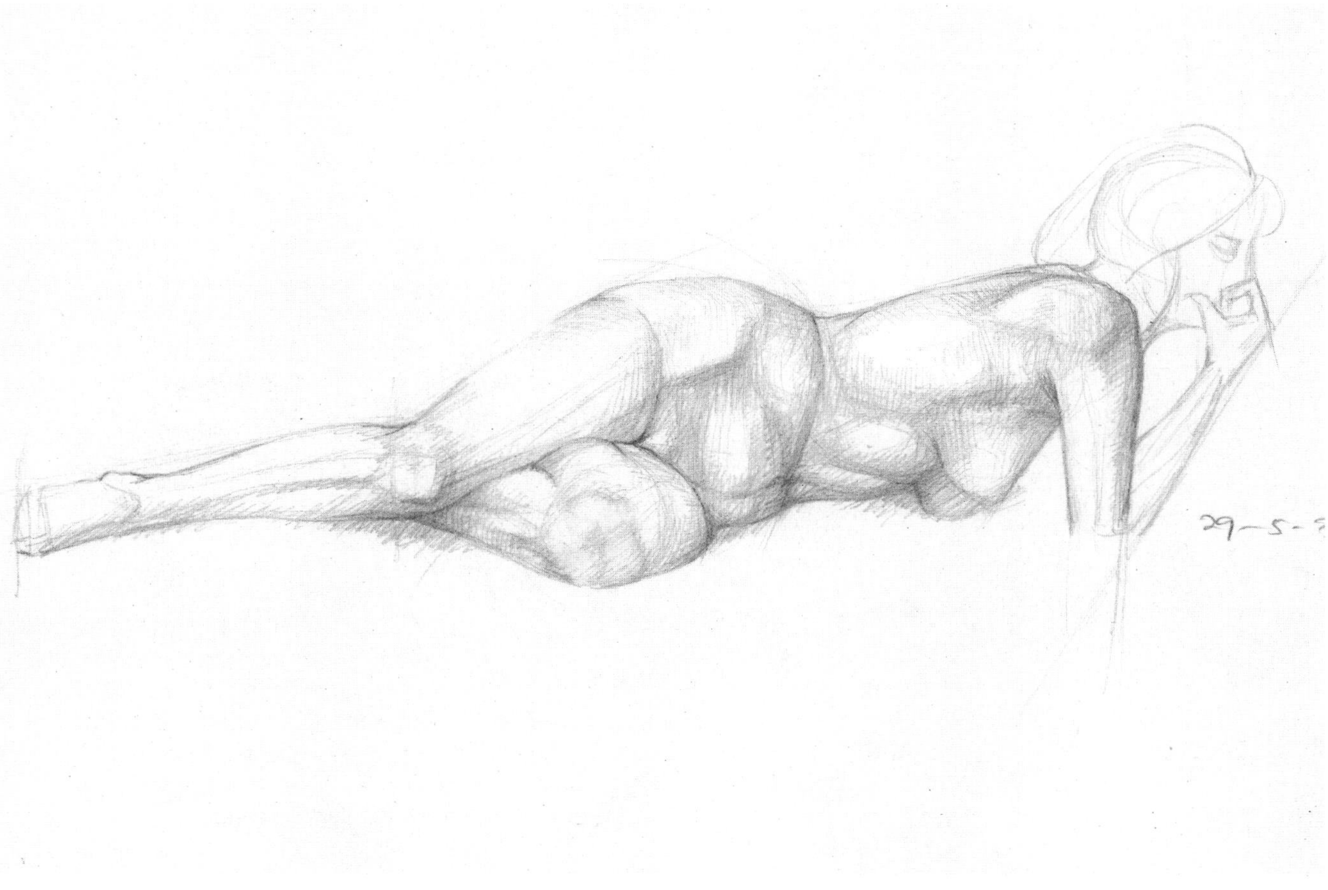

192 MULTIPLE FUNCTIONS IN SITTING AND THEIR CONSEQUENCES

The main points of emphasis in this study are the columnar character of the supporting arm and the resulting one-sided drop of the shoulder girdle, the unequal loading of the seated weight and consequent compression and twisting of the abdominal wall.

Student of painting/graphic art, fourth semester

193 DISTORTIONS AND TENSIONS IN A LYING POSE INVOLVING MOVEMENT

The pelvis on its side stands up while the rib cage is flat on its back; the arms clasped behind the head cause a sunken, twisted tension in the abdominal wall, a flattening of the breasts and a directional pull on the neck (as a result of the narrow attachment to the extended large pectoral muscle).

Student of painting/graphic art, fourth semester

194 GIVING SHAPE TO FREELY INVENTED MOVEMENTS

It is an important recuperative phase after intense study from life to experiment imaginatively with the forms the body takes up in invented gestures. Even if quite a few anatomical factors are incorrect, the student's powers of visualization are working.

Student of stage design, fourth semester

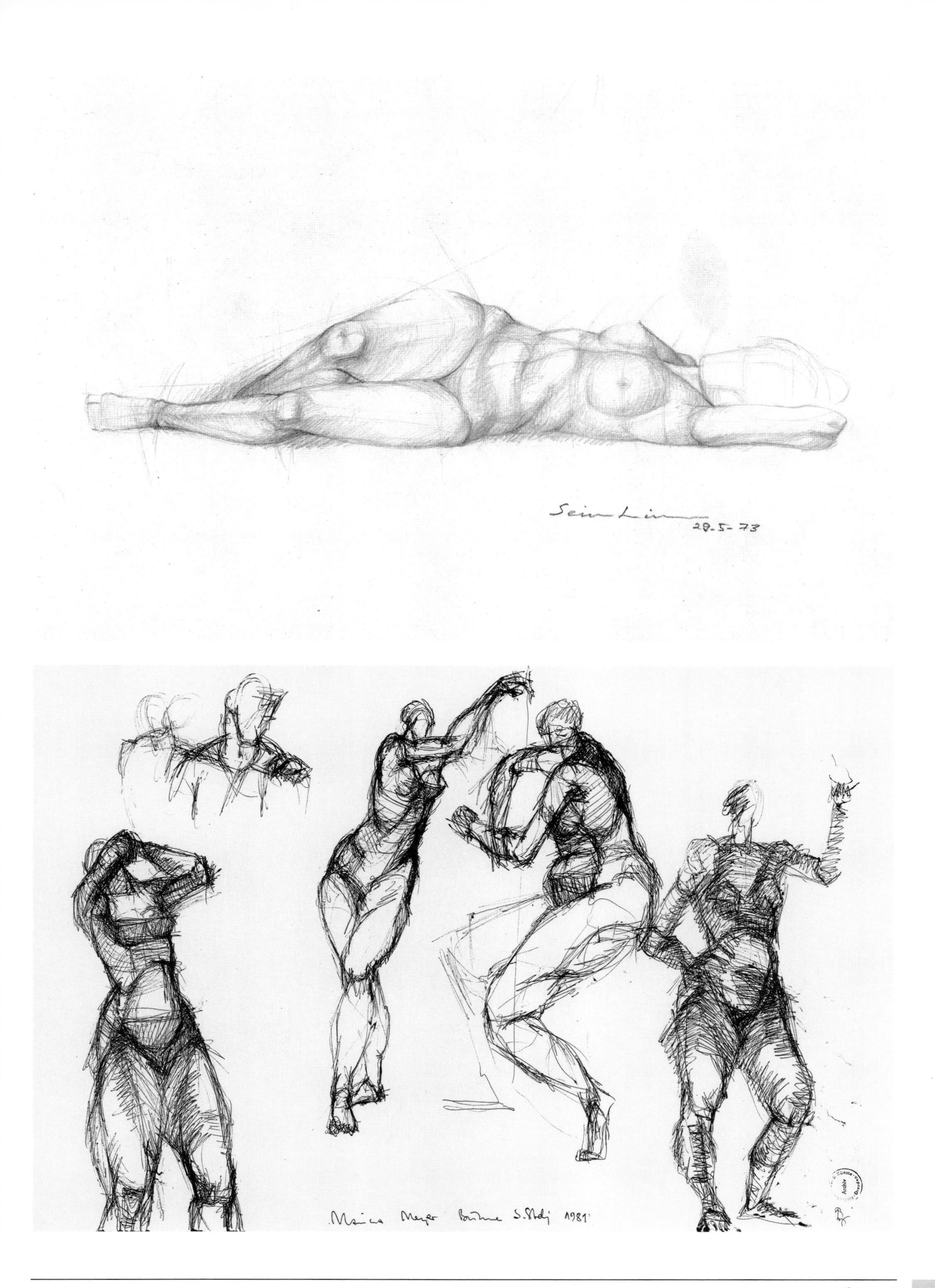

195 FREE EXECUTION OF A GREATLY FORESHORTENED LYING POSE

The graphically free expression of something seen and experienced is even more unreserved than in fig. 188. The powerful foreshortening is chosen in such a way that the stomach can still function as a necessary intermediary form between the shoulders and thighs, so enabling us to perceive the figure correctly. The areas of hatching enclosing the figure create a concise environment putting it into context.

Student of painting/graphic art, end of fourth semester

196 CONSISTENT CONSIDERATION OF THE VIEWING ANGLE IN CONSTRUCTING FORMS

The overall three-dimensional situation shows the figure in a slanting position running from the front elbow to the feet, with the model looking slightly upward. The student has followed through the modeling of the forms with complete sequential correctness using visualization of the basic facts that cylindrical and rectangular forms and cones are projecting toward or receding from the viewer, and of the ways in which they do so.

Student of painting/graphic art, fourth semester